Platform Regulation

Platform Regulation

Exemplars, Approaches, and Solutions

PRADIP NINAN THOMAS

OXFORD
UNIVERSITY PRESS

Great Clarendon Street, Oxford, ox2 6dp,
United Kingdom

Oxford University Press is a department of the University of Oxford.
It furthers the University's objective of excellence in research, scholarship,
and education by publishing worldwide. Oxford is a registered trade mark of
Oxford University Press in the UK and in certain other countries

First Edition published in 2023

Impression: 1

Published in the United States of America by Oxford University Press
198 Madison Avenue, New York, NY 10016, United States of America

British Library Cataloguing in Publication Data

Data available

Library of Congress Control Number: 2022913664

ISBN 978-0-19-288796-2

DOI: 10.1093/oso/9780192887962.001.0001

Printed and bound in India by Replika Press Pvt. Ltd.

*To All Creatures Great and Small that have kept me Company as I wrote
this book. Thankyou.*

*The Tree Snake and the Blue Tongue,
Magpie and Kookaburra,
To Kamikaze Peewees,
And
Moss our Goldie
Gentle Soul
Getting Rickety Too Soon*

Preface

Over the last decade or two, a handful of powerful, monopolist platforms have embraced our lives, intermediate our socialities and relationships (Facebook), what we search for on the Internet from news to menus and prices of products (Google) and our online purchases (Amazon). We are living in a global economy that is fuelled by the monetization of affect and we are now only too aware that these platforms are very systematically using the advantages stemming from algorithmic power and platform externalities to mine and privatize personal data that is in turn sold to advertisers who target not just the present but also future the economic behaviours of its users. We are also now aware of the complicity of some of these platforms in data breaches that have contributed to the making and unmaking of political fortunes of key political parties in the United States and the outcome of the Brexit vote in the United Kingdom. Platforms, as Benjamin Bratton (2015)[1] has argued, occupy an increasingly important space in planetary-scale computation, are an essential aspect of the Stack, the layered systems of hardware and software, their many material and immaterial expressions, and their protocols that each one of us in a connected world is involved in negotiating on a daily basis. Shoshana Zuboff (2019:11),[2], however goes behind the Stack as it were, and focusing on Google in particular has argued in her volume *The Age of Surveillance Capitalism* that there is nothing that is benign about platforms that are at the core of what she has called 'surveillance capitalism' 'Surveillance capitalism operates through unprecedented asymmetries in knowledge and the power that accrues to knowledge. Surveillance capitalists know everything *about us*, whereas their operations are designed to be unknowable *to us*. They accumulate vast domains of new knowledge

---

[1] Bratton, B. H. (2015), The Stack: On Software and Sovereignty, MIT Press, Cambridge: Massachusetts and London: England.
[2] Zuboff, S. (2019), The Age of Surveillance Capitalism: The Fight for a Human Future at the New Frontier of Power, Profile Books, London.

from us but not *for us*'. There is, from the point of view of Zuboff, nothing that is accidental about the Stack.

This unprecedented power of platforms is, however, being challenged today. Data breaches, evidence of platform manipulations, platform complicities with state surveillance, their monopolist behaviours, and its consequences for competition and data privacy have become the basis for regulatory responses from governments throughout the world. National and regional courts of law have collected masses of evidence on myriad forms of platform illegalities that discriminate against competitors and that point to the privatization of personal data on a global scale. The fact that the world's largest platforms are of US origin may just be coincidental although in the context of the European Union (EU) and their commitment to a single European Digital Market, their historical antipathy towards US-based information and cultural imperialism, now includes US-based data imperialism. After what one can consider was an extended 'grace period' during which platforms were able to constantly and systematically innovate in very disruptive ways in environments that were devoid of any regulatory oversight, it would seem that the goodwill that governments had towards platforms is no longer *carte blanche* but is based on a critical, new-found discernment of algorithmic fiat and the role played by these platforms in shaping individual behaviours, political choices, and economic futures. The default position was self-regulation— a stance that Big Tech continues to believe in as the only feasible solution. However, self-regulation has largely failed and there is a widely held view that it has not been able to deal effectively with some of the fundamental issues related to Big Tech companies—the fact that they exercise effective control over both personal and non-personal data and that they are inherently anti-competitive in their economic behaviour. At the very same time, platform entanglements and platform compromises with authoritarian governments and political parties have highlighted the fact that regulation can be shaped by extra-regulatory factors, such as religious nationalism and reasons of 'national security' as is the case in India, and systematically used by political parties such as the Partido Federal ng Pilipinas and the current President of the Philippines, BongBong Marcos to rewrite the history of corruption and misrule by his father Ferdinand Marcos in the Philippines. These witting or unwitting, tacit or open collaborations between platforms and political parties that have benefited greatly from platform dispensations, exceptionalisms, and arrangements have muddied the waters as it were. As a result, the regulation of platforms in jurisdictions around the world is bound to be complex given the nature of the deep entanglements that exist between platforms and the nation-state on the one hand and platforms and citizens on the other. While platforms such as Facebook/Meta have their own oversight body that is involved in content moderation, as Robert Gorwa (2019:2)[3] has argued

> The current 'platform governance' status quo—understood as the set of legal, political, and economic relationships structuring interactions between users, technology companies, governments, and other key stakeholders in the platform ecosystem …—is rapidly moving away from an industry self-regulatory model and towards increased government intervention.

This volume highlights four regulatory responses from four jurisdictions—namely the EU, the United States, India, and Australia. The reason for choosing these four case studies and not others is because of my own research interest in the case of regulation in India and the EU, teaching platform regulation on a course offered at the Indian Institute of Technology (Bombay) during 2020–2021, and also because the EU, the United States, and Australian cases of regulation are topical and have fed into global discussions across multiple media sectors. Facebook's very public tussle with regulatory authorities and the government of Australia, in particular, its blocking of news has been trending on news platforms across the world. As a researcher primarily interested in the political economy of communications, I have, over the years, written extensively on the many ways in which the digital has been shaped by structures and institutions, by political imperatives and geopolitical considerations, by the power of Big Tech and the State—to create on a global scale a digital mode of production that has, to some extent, subsumed other extant modes of production in its image.[4] I have also written about

---

[3] Gorwa, R. (2019), The platform governance triangle: Conceptualising the information regulation of online content (1–22), Internet Policy Review, 8 (2).

[4] Thomas, P. N. (2012), Digital India: Understanding Information, Communication and Social Change, Sage, New Delhi; The Politics of Digital India: Between Local Compulsions and Transnational Pressures (2019), Oxford University Press, New Delhi; Information Infrastructures in India: The Long View (2022), Oxford University Press, New Delhi.

and reflected on the consequences of this digital imperative for the public good. Arguably platform capitalism has contributed to a sharpening of the contradictions in capitalist value creation in a transactional economy in which the conditions of labour and labouring have become precarious for millions of workers. This volume highlights the power of Big Tech but critically the counter power of the State to claw back some advantage in shaping the terms of Big Tech's presence and operations in key jurisdictions around the world—platform regulation contributing to a level playing field and competition, to strengthening the exchequer and the rights of consumer-citizens.

This story of regulation is incipient and is therefore incomplete. Regulation is progressing in fits and starts in different jurisdictions around the world. And as such, the case studies in this volume are by no means a reflection of wholly complete processes but are rather, snapshots of regulation in process at a particular moment in time. In some jurisdictions such as the EU, pathways are clearer. Without a doubt, the response from the EU via instruments such as the General Data Protection Regulation (GDPR, 2018) and Digital Services Act and Digital Markets Act (2020) communicate a clear and unequivocal intent to regulate Big tech and the EU has as it were, thrown down the gauntlet to global platforms. Many of these platforms have been taken to court and multiple anti-trust rulings have highlighted in no uncertain terms, the EU's vision of a digital society that is based on the public good. There are of course critics who point to the unworkableness and complexities of principles such as the 'right to be forgotten' although arguably, this can be seen as a counter-disruptor to platform fallacies and futures based on their perfect control over personal data in perpetuity. The planetary scale of platform expansion across multiple sectors has been called into question and there is a renewed call from governments around the world, even from the United States, to regulate platforms that have colonized the search, social, and e-commerce markets and shaped it in their own image.

Platform regulation is not going to follow an ordered, readymade script but will be shaped by local exigencies, local politics, and by ideas such as data nationalism. In the United States, and despite a flurry of actions taken by the Federal Communications Commission (FCC) and Federal Trade Commission (FTC) and bipartisan interventions, the need for platform regulation continues to be debated and contested, whereas in the EU, regulatory frameworks are currently being established. From the perspective of an outsider looking in, regulation is, at the moment at least, a messy affair and there are no guaranteed outcomes. Not surprisingly, these attempts to regulate are being contested by platforms that have vast amounts of capital at their disposal. The EU's Digital Markets Act that legitimizes competition has been taken to court by 'gatekeeper' platforms including Apple and Google[5] There are academic supporters of Big Tech who favour competition but not government regulatory oversight.[6] The platforms do have a lot to lose if plans to curb their activities, levy taxes, pay for public journalism as is the case in Australia, break them up or turn them into public utilities become a reality. There are bound to be regulatory failures and mishaps and government retreat from their public support for data nationalism and data privacy legislations although also occasionally, there will also be governments advancing thought-through, manageable models for regulation supported by investments in regulatory personnel and infrastructure, the harmonization of regulatory laws and effective enforcement. There have been some very interesting discussions on related issues such as the economic governance of data by the Bengaluru-based NGO, IT for Change. Singh and Gurumurthy (2021)[7] have contrasted the EU's, largely individual-oriented approach to data governance to the community-based framework articulated in a draft report of India's Committee on Data Governance Framework. They define community data governance as follows 'Community data governance refers to the norms, institutions and processes that determine how power and responsibilities over community data are exercised, how decisions are taken, and how citizens—people and communities—and businesses participate in and benefit from the management of community data' (30). While such pre-emptive policy interventions are important, how to operationalize a community-based data sharing approach or for that matter

---

[5] Mellor, S. (2022), Apple and Google criticize the new EU Digital Markets Act that will radically change the way they have operated for the past twenty years, Fortune, 26 March. Available at: https://fortune.com/2022/03/25/apple-google-criticize-eu-digital-markets-act/

[6] Lemley, M. A. (2021), The contradictions of platform regulation (304–336), Journal of Free Speech Law, 1 (303).

[7] Singh, P. J. & Gurumurthy, A. (2021), Economic Governance of Data: Balancing Individualist-Property Approaches with a Community Rights Framework (1–32), IT for Change, Bengaluru.

a mixed semi-commons approach to data sharing[8] in a complex country such as India with its social, cultural, and economic hierarchies within an explicitly religious nationalist state remains a challenge.

The October 2021 agreement on reform of the international tax system by all states belonging to the Organisation for Economic Co-operation and Development (OECD)[9] and formally ratified by the G20 that met in Rome in November 2021, mandates a minimum 15% tax rate for multinational companies with annual revenues of $890 million (the Global Anti-Base Erosion Rule—GloBE, Pillar 2) while also re-allocating some taxing rights to the jurisdictions in which these MNCs operate in—Pillar 2 (see the OECD report (2021, Two Pillar Solution). The OECD has maintained that MNCs have taken advantage of a diversity of tax regimes, paid taxes in countries that have low corporate tax rates such as Ireland thus depriving countries of $100–240 billion in tax revenues (Dangor 2021).[10] While this is a major step towards the establishment of a global tax regime fit for a global digital economy, its proposed implementation in 2023 is bound to be contested by MNCs, and by nation states that already have in place unilateral digital taxation regimes, however imperfect, such as the Equalisation Levy in India.

So to a large extent, what we are witnessing in terms of platform regulation is just the beginning of a protracted struggle that will be waged in many jurisdictions around the world to bring platforms to heel. These moves will be met by counter-moves by platforms that will use their lobbies and money power to bypass, derail, compromise, and co-opt regulation. The history of anti-trust actions against Big Tech in the past has been long drawn out and for the most part unsuccessful. The anti-trust case against IBM, for example, began in 1969, dragged on for 13 years and involved 974 witnesses, 104,400 pages of transcript, 200 concurrent

---

[8] Gurumurthy, A. & Chami, N. (2022), Governing the Resource of Data: To What End and for Whom? Conceptual Building Blocks of a Semi-commons Approach (1–28), Data Governance Network, IDFC Institute, Data Governance Networking Paper 23.

[9] OECD/G20 Base Erosion and Profit Shifting Project: Two Pillar Solution to address the tax challenges from the Digitalisation of the Economy (2021), October (1–22). Available at: https://www.oecd.org/tax/beps/brochure-two-pillar-solution-to-address-the-tax-challenges-arising-from-the-digitalisation-of-the-economy-october-2021.pdf

[10] Dangor, G. (2021), G20 signs off on 15% global minimum corporate tax—Here's how it will work, Forbes, 11 July. Available at: https://www.forbes.com/sites/graisondangor/2021/07/11/g20-signs-off-on-15-global-minimum-corporate-tax-heres-how-it-will-work/?sh=3f54c79b1c7e

lawyers, and cost millions of dollars (Villasenor 2020)[11] that ultimately did not amount to much. We simply have to recognize the fact that platform regulation has come to the forefront of public agendas precisely because Big Tech were allowed to acquire companies that contributed to their becoming expansive monopolies—Facebook's acquisition of Instagram in 2012 for $1 billion and WhatsApp in 2014 for $22 billion, Google's acquisition of Waze in 2013 for $966 million, YouTube in 2006 for $1.6 billion and Fitbit in 2021 for $2.1 billion and Amazon's take-over of Whole Foods in 2017 for $13.7 billion—a total of 770 acquisitions by Big Tech over three decades (CBInsights 2020).[12]

There is no denying of the fact that there are governments that have benefited immensely from platform complicities and that continue to benefit from a given platform's working arrangements in return for regulation-lite business environments. The attempts by Facebook to shut down the page and Instagram accounts of the *Kisan Ekta Morcha*, the official digital voice of farmers demonstrating in North India clearly points to government pressure to cut off a vital source of updates on what is an extraordinary movement against the implementation of three farm laws that farmers argue would result in the accentuation of the privatization of agriculture (Indian Express 2020).[13] Cutting off a vital source of information is like cutting off access to public utilities such as water and electricity and so it is little wonder that there are those who have argued that platforms need to be regulated like public utilities because their services have become vital to the quality of lives that people lead. Access to information, it has been argued, is an essential human right that is on par with access to other basic rights such as food, employment, and shelter. This would indicate of course that platforms have become vital to the business of everyday life—our social life, how and who we connect to, what we buy online, and what we search for as we navigate content online that makes a

[11] Villasenor, M. (2020), Is global antitrust upto the challenge of Big Tech?, Council on Foreign Relations, 7 December. Available at: https://www.cfr.org/blog/global-antitrust-challenge-big-tech

[12] Visualizing tech giants billion-dollar acquisitions (2020), CBInsights: Research Briefs, 5 May. Available at: https://www.cbinsights.com/research/tech-giants-billion-dollar-acquisitions-infographic/

[13] Social media crucial for our agitation: Farmers on Facebook shutting down page (2020), The Indian Express, 21 December. Available at: https://www.newindianexpress.com/nation/2020/dec/21/social-media-crucial-for-our-agitation-farmers-on-facebook-shutting-down-page-2239249.html

difference in our lives. This points to the fact that platforms by their very nature are multi-sided and have come to their own in thousands of different contexts around the world—from social movements and the Arab Spring to corporate boardrooms and the everyday practices of domestic workers and daily wage labourers. Be that as it may, we recognize that this vast, transactional economy is fuelled by the monetization of affect—and it is this business model of platforms that has come under scrutiny in the light of the many scandals linked to data manipulations and data transfers.

In the absence of global regulatory standards, blueprints, and pathways, it is going to be nigh impossible for economically weak countries to bargain with or restrict the activities of platforms. One question is whether there is a need for a new body or supranational organization such as the OECD to establish models and frameworks for platform regulation. A key issue for Big Tech is how to respond to data nationalism and data localization and a strategy employed could well be directed towards assuaging data concerns, allaying the fears of governments, and finding win-win situations based on attractive data retention *and* data portability options.

Platform regulation is a complex issue precisely because it overlaps with other aspects of digital regulation. Internet governance of course provides the larger framework for platform governance. Today there are any number of very specific, discrete legislations aimed at curbing Net-based money laundering, the use of the Deep Net by terrorist organizations and paedophile networks although one can argue that a plethora of issue-specific legislations does not amount to a comprehensive regulatory strategy for the Internet. Governance and regulation although related concepts mean different things. While governance is linked to systematic, rule-making, regulation is about the establishment of legal frameworks and the boundaries (in our case) for platform behaviour accompanied by legal sanctions and enforcement. Regulation can set the terms for platform power. If regulators in the EU have their way, Google and Facebook's vast holdings as global conglomerates can, in the worst-case scenario, be dismembered into bite-sized operations suitable for a competitive marketplace.

An integrated approach to platform regulation will take time and it will be based on an incremental approach, a brick-by-brick process although this task will be hugely complicated by the fact that platforms will continue to innovate and develop. Regulation will not take place in a vacuum but will be accompanied by more innovation-based digital disruptions of cultures, habits, of standard ways of doing things, and ways of seeing the world. This remains a conundrum for regulators—what to regulate as a priority in the context of a fast-changing technological environment, what to regulate in the interests of the public good and what not to regulate in the interests of maintaining market dynamism and digital innovation? Then there is the question to do with the processes of regulation and whether or not a range of stakeholders should be involved in platform regulation. I have in this volume focussed on the role played by the state and regional players such as the EU in platform regulation. Civil society has of course a lot to say on regulation although representatives from civil society are by no means accepted as equal partners or as stakeholders by most governments. The EU remains an exception to this rule although even in this case, it is unclear as to the extent to which contributions from civil society have made a difference in policy making. I have not therefore explored in any detail the issue of whether or not civil society ought to be involved in the regulation of platforms although I have highlighted some civil society actions related to Big Tech in the context of the EU in the concluding chapter. The role of civil society is an added complexity although are variations across jurisdictions on the scope, consistency, presence, and their ability to influence policy on platform regulation for a just society.

In spite of these limitations, I hope that this volume provides an introduction to some of the issues and challenges related to platform regulation, the conundrums, and paradoxes but also to some of the well-conceived and manageable regulatory pathways currently being explored by national and regional governments. If there is a key takeaway from this volume, it is the fact that regulation is just too important to be left to the whims and fancies of platforms. We simply have to move beyond excuses for the status quo and need to contest the threats from Big Tech to determine the terms for regulation. In a world that is increasingly becoming aware of the reality of multiple risks—an important objective of platform regulation is to limit the risks associated with ultra-powerful platforms and advance the common good.

Contents

1

Platform Regulation

Three Reasons

In the space of less than two decades, the global capitalist economy is today critically fuelled by data generated within a platforms-based mode of production. I am not suggesting that earlier modes of production—agriculture and manufacturing, for example, are no longer relevant. They are, although these earlier modes of production are steadily being embraced by the logic of platforms and platform capitalism. On the one hand, there is nothing extraordinary about this turn since capitalism's forte has been its ability to ceaselessly look for new sources of value in the aftermath of economic recessions and falling value that led to the rise of the services economy in the 1970s[1] and in the aftermath of the dotcoms bust in 2001–2002. There was a fundamental restructuring of labour involved in the generation of value from intangible goods and services, a process that has accentuated in the context of the globalization of the gig economy. On the other hand, the platform economy's focus on data-based value generation has resulted in data—all data, personal and impersonal, becoming the basis for value generation resulting in the logic of data becoming the source of value across all productive sectors in an economy that increasingly is being charged by literally billions of data transactions in real time.

This book is not about platform capitalism, its logic or key players that a number of academics including Nick Srnicek[2] and van Dijk[3] have very fulsomely described. Rather, it is about contemporary moves being made by nation-states such as the United States, Australia, and India, and

---

[1] Walker, R. A. (1985), Is there a service economy? The changing capitalist division of labour (42–83), Science & Society, 49 (1).

[2] Srnicek, N. (2017), Platform Capitalism, Polity Press, Cambridge & Malden.

[3] Van Dijk, J. (2013), The Culture of Connectivity: A Critical History of Social Media, Oxford University Press, New York.

Platform Regulation. Pradip Ninan Thomas, Oxford University Press. © Pradip Ninan Thomas 2023.
DOI: 10.1093/oso/9780192887962.003.0001

regional entities such as the European Union (EU) to curb the cultural, social, political, and economic influence of a handful of mega-platform behemoths, namely Google, Facebook, and Amazon through regulating their operations, taxing the economic value that they generate and establishing a level playing field for new entrants and competition in the social media, search, and online retail markets. It is their monopoly or at best duopoly position, in markets across the world and their ability as incumbents to 'game' the market and restrict competition that has been a key reason for the increasing interest shown by nation-states to regulate platforms. However, it is not just their economic stranglehold but also the ability of social media platforms to 'game' the terms of electoral discourse and political fortunes in nation-states following the election of Trump, Brexit, Modi, and perhaps most egregiously, their monetization of, control over and shaping the 'affective' lives and behaviours of citizen-consumers and 'rhetorical' control that governments are most concerned with. This concern, to some extent, relates to the exercise of power over citizens that had hitherto been exercised by governments as elected representatives but that, to some extent, has been usurped by unelected transnational corporations that had no fixed address but operated in jurisdictions that they had little or no allegiance to. With the benefit of hindsight, this regulatory moment is a consequence of the dereliction by the neo-liberal State of their responsibility to regulate the online economy in line with the offline economy—a dereliction that was fundamentally a result of their absolute commitment to economic growth at the expense of shaping the terms and drivers of that growth. It has taken multiple platform scandals for governments to act on regulation.

While the regulation of platforms has its own intrinsic merits from both the perspectives of governance and the economy, one of the issues that needs to be foregrounded is the fact that the imperative behind platform competition is to a large extent being shaped by a high-stakes game played by the EU and China in particular against US-based incumbents who have been able to capitalize on network externalities and build powerful global empires in the domains of search and sociality. While it is necessary to acknowledge the 'progressive' nature of the EU's General Directive on Platform Regulation (GDPR), especially its inclusion of novel but complex concerns such as the 'Right to be Forgotten', the EU's contestation of US digital imperialism has been accompanied by its own imperialism directed against countries in the developing world. This has taken the form of the insertion of clauses in bilateral trade agreements and free trade negotiations that explicitly contradict the EU's own moral reasonings used to counter US platform hegemonies. These clauses on digital trade highlight EU exceptionalism and the right to either extractive practices or prohibitions in areas such as cross-border data transfers, data localization, local data processing, non-disclosure of the source code of software and related algorithms, and customs duties on digital goods and services. As Scasserra and Elebi (2021:1) have observed the 'EU has adopted a colonialist strategy, going out to hunt for data from the global South, in order to position its own companies in the new global cybernetic value chains. To empower its own Big Tech, the EU is seeking to force through clauses in trade negotiations that will hinder digital industrialisation, restrict necessary state oversight of corporations and undermine citizens' rights elsewhere, in particular in developing countries'.[4]

So what are some of the characteristics of platforms that regulatory regimes are keen to control and what are the ecosystems that are now essential to platform performances? van Dijk's (2012:28)[5] characterization of platforms as both *techno-cultural constructs* and *socio-economic structures* provides us with the necessary scaffolding to understand platforms as material actants consisting of hardware, software, and networks along with shared technologies and interfaces that can be built upon and used by third parties to build more platforms. Platforms beget platforms.[6] And it is this proliferation of platforms that is the basis for the platform economy and its constitutive role in the contemporary global economy. However, a handful of transnational platforms have clearly begun to exert hegemonic control over platform architectures and shared interfaces to create their dominance over search, sociality, and online commerce with significant consequences for those platforms that depend on the incumbents. Arguably, Facebook/Meta and Google have become

---

[4] Scasserra, S. & Elebi, C. M. (2021), Digital Colonialism: Analysis of Europe's Trade Agenda (1–50), Trade and Investment Policy Briefing, Transnational Institute, Amsterdam, October.

[5] Van Dijk, J. (2013), The Culture of Connectivity: A Critical History of Social Media, Oxford University Press, New York.

[6] Kenney, M. & Zysman, J. (2016), The rise of the platform economy (61–69), Issues in Science & Technology, 32 (3).

critical infrastructures since millions of businesses around the world are dependent on the business models aligned with these platforms. Just as consumers and businesses are hard hit when submarine cables that carry intercountry data are damaged, a Facebook outage can result in massive losses to businesses. So it is this advantage as incumbents to control, curate, and monetize the generation, flows, and consumption of data in their role as mediators that has given them unparalleled, global power. While comparisons have been made with global monopolies of yesteryear including Standard Oil and AT&T, the influence of these new hegemons is different precisely because that influence works through 'connectivity' and therefore includes all those who are connected—businesses, consumers, citizens, and all productive sectors in society that use, and have now become dependent on these platforms.

There are broadly speaking three sets of reasons for regulating global platforms. The first one is internal to the platforms themselves and the perceived use/misuse of and control cover algorithmic power, machine intelligence and their applications in the 'deep reading' of processes, and predictive forecasting of sentiments and affect, behaviours. There is a broad consensus among governments that control over the operations of code bestows key global platforms unprecedented power over consumers/citizens, their Voice, expression, and socialities and political processes, accompanied by concerns related to the privacy of personal and non-personal data and the collection of data on publics that hitherto was the preserve of the State. The second, has to do with the consequences of incumbency and, in particular, its effects on competition in markets that are dominated by monopolies or at best, duopolies. And third, the need for a harmonization of a global tax system that applies specifically to global platforms that have largely been able to evade paying taxes proportionate with their earnings and paying taxes in the jurisdictions that they operate in. This harmonization will potentially result in major financial dividends for governments in the forms of taxes. The October 2021 agreement between G7/G20/and OECD members to a 15% minimum corporate tax rate applicable to all MNCs (Thomas 2021)[7] and data firms involved in online advertising and marketplaces, online search engines,

---

[7] Thomas, D. (2021), Nations agree to minimum 15% corporate tax rate, BBC News. Available at: https://www.bbc.com/news/business-58847328

social media platforms, digital content services, cloud computing services, sales from user data, digital intermediary services will be based on a tax rate that will specifically apply to data firms that annually earn a minimum of US$867 million. The OECD that spearheaded this reform estimates that $125 billion will be reallocated to governments from close to 100 of the world's largest MNCs as a result of this agreement.[8]

Algorithmic Governance and Its Regulation

Rouvroy and Berns (2013:167–169)[9] describe three stages of algorithmic governmentality: (1) the collection of large amounts of personal and non-personal data and their storage, (2) data processing and knowledge production based on machine learning that generates correlational knowledge that becomes the basis for norms, and the (3) application of this statistical knowledge to anticipate human behaviour linked to profiles. This fundamentally is about ensuring predictive effectiveness, the anticipation of behaviour based on a constant evaluation of personal data resulting in the creation of profiles of the perfectly decontextualized individual shorn of any individuality and that can, in turn, be commodified. Rouvroy and Berns use the term 'Algorithmic governmentality to refer very broadly to a certain type of (a) normative or (a) political rationality founded on the automated collection, aggregation and analysis of big data so as to model, anticipate and pre-emptively affect possible behaviours' (170). The programmed sociality of Facebook is based on algorithmic profiling of its customers and their blackboxing, and this pre-emptive profiling is highlighted in the example of the machine learning algorithm supplied by the firm Geofeedia to the Baltimore Police Department after the custodial death of an Afro-American, Freddie Gray. Analysing feeds from Facebook, Instagram, YouTube, and other sites on planned protests, 'terabytes of images, video, audio, text, and biometric and geospatial data

[8] Kvetenadse, T. (2021), 136 countries agree to establish minimum 15% corporate tax rate, Forbes, 8 October. Available at: https://www.forbes.com/sites/teakvetenadze/2021/10/08/136-countries-agree-to-establish-15-minimum-corporate-tax-rate/?sh=b464f7564bbe

[9] Rouvroy, A. & Berns, T. (2013), Algorithmic governmentality and prospects of emancipation: Disparateness as a precondition for individuation through relationships? (163–196), Reseaux, 177 (1).

from the protests of the people of Baltimore were rendered as inputs to the deep learning algorithms' leading to the arrests of 49 children because the algorithm had highlighted them as high risk, potential agitators[10] And in the case of Google, influencing the algorithm through a targeted avalanche of biased search terms, has been a strategy used by right-wing groups to create 'trustworthiness'[11]—a strategy that has been used by Holocaust deniers and the alt-right in the United States, and that has been enhanced by the paid ranking of search.[12]

As Bucher (2018:4)[13] explains, 'In ranking, classifying, sorting, predicting, and processing data, algorithms are political in the sense that they help to make the world appear in certain ways rather than others. Speaking of algorithmic politics in this sense, then, refers to the idea that realities are never given but brought into being and actualized in and through algorithmic systems. In analyzing power and politics, we need to be attentive of the way in which some realities are always strengthened while others are weakened, and to recognize the vital role of non-humans in co-creating these ways of being in the world … algorithmic power and politics is neither about algorithms determining how the social world is fabricated nor about *what* algorithms do per se. Rather it is about *how* … and *when* different aspects of algorithms and the algorithmic become available to specific actors, under what circumstance, and who or what gets to be part of how algorithms are defined'. In other words, machine-based decision-making based on artificial intelligence (AI) can have real-life consequences for people. Algorithmic biases, perfected by profiling technologies can affect their credit worthiness, hiring, access to services, and contribute to their perpetual criminalization and stereotyping.

This algorithmic mode of production, and in particular algorithmic sorting and classifying of people and the gaming of code, can and does

---

[10] Amoore, L. (2020), Cloud Ethics: Algorithms and the Attributes of Ourselves and Others, Duke University Press, Durham.

[11] Solon, O. & Levin, S. (2016), How Google's search algorithm spreads false information with a rightwing bias. The Guardian, 16 December. Available at: https://www.theguardian.com/technology/2016/dec/16/google-autocomplete-rightwing-bias-algorithm-political-propaganda

[12] Bradshaw, S. (2019), Disinformation optimised: Gaming search engine algorithms to amplify junk news (1–24), Internet Policy Review, 8 (4). Available at: https://policyreview.info/pdf/policyreview-2019-4-1442.pdf

[13] Bucher, T. (2018), If … Then: Algorithmic Power and Politics, Oxford University Press, New York.

result in the creation of discriminatory marketplaces and in the perpetuation of pre-existing divides. Algorithms used for example in e-government, need to be trialled extensively before their roll out and governments need to have robust testing procedures to check for algorithmic biases. In the United Kingdom, a number of councils have had to withdraw algorithms used to assess benefit and welfare claims—actions that were hastened after an algorithm used by the exam regulator *Ofqual* downgraded 40% of A-Level grades assessed by teachers in 2020.[14]The Committee on Standards in Public Life, United Kingdom published a review of AI and Public Standards[15] that included 15 recommendations including the following: 'Recommendation 9: Evaluating risks to public standards—Providers of public services, both public and private, should assess the potential impact of a proposed AI system on public standards at project design stage, and ensure that the design of the system mitigates any standards risks identified' along with other recommendations including:

- greater transparency by public bodies in use of algorithms,
- new guidance to ensure algorithmic decision-making abides by equalities law,
- the creation of a single coherent regulatory framework to govern this area,
- the formation of a body to advise existing regulators on relevant issues,
- and proper routes of redress for citizens who feel decisions are unfair.[16]

Algorithmic governance arguably is a form of control, that, is to a large extent today being exercised by privately owned platforms and one of the key questions is whether firms associated with the commercial uses of

[14] Marsh, A. (2020), Councils scrapping use of algorithms in benefit and welfare decisions, The Guardian, 24 August. Available at: https://www.theguardian.com/society/2020/aug/24/councils-scrapping-algorithms-benefit-welfare-decisions-concerns-bias

[15] Artificial Intelligence and Public Standards: A Review by the Committee on Standards in Public Life (2020), The Committee on Standards in Public Life, March, 1–77, London.

[16] Clement-Jones, T. (2021), Tackling the algorithm in the public sector, The Constitution Society, 19 March. Available at: https://consoc.org.uk/tackling-the-algorithm-in-the-public-sector/

personal data should be placed under regulatory oversight. To governments around the world, the unprecedented power of a handful of corporations to shape economic and political behaviours and set the terms for 'agency', to include and exclude and to optimize a very precise, focussed, and calibrated predictive monetization of consumers is a cause for concern[17]—hence the turn towards the governance of algorithms and the regulation of AI.[18] Governments have of course been deeply involved in the use of social media platforms for advancing their own political objectives although it is clear from Twitter's relationship with the former President of the United States, Donald Trump, that their power to include Alt-right views, draw the boundaries of discourse with respect to matters such as hate speech and suspend Trump's account when it was perceived to cross that line, demonstrated the power of a handful of platforms to police, some would say very belatedly and perfunctorily, the terms of online political discourse. If the Twitter handles of Trump could be suspended, so could that of other politicians, influencers, and celebrities. It is in the EU that the regulation of AI and algorithms has received considered responses. The European Parliament's 'A Governance Framework for Algorithmic Accountability' (2019)[19] Section 3.9.6 on State Intervention explores the measures that can be implemented including command and control regulation based on legislation, regulatory bodies involved in standards setting involved in setting performance, design, and liability standards, Light Regulation and Hard Regulation—options that are applicable to both the public and private sectors (45–50). The EU's Digital Services Act and Digital Markets Act offer guidelines for platform accountability—the culmination of a variety of actions by the EU including enforcing transparency with respect to Google, Amazon, and Microsoft ranking of online search, paid ranking, and ranking of products and services owned by the search companies over others.[20]

---

[17] Katzenbach, C. & Ulbricht, L. (2019), Algorithmic governance (1–18), Internet Policy Review, 8 (4).

[18] Ebers, M. & Gamito, M. C. (2021), Algorithmic Governance and Governance of Algorithms: Legal and Ethical Challenges, Springer Nature, Cham, Switzerland.

[19] A governance framework for algorithmic accountability and transparency (1–124), European Parliament, Study Panel for the Future of Science & Society, Scientific Foresight Unit, Brussels. Available at: https://www.europarl.europa.eu/RegData/etudes/STUD/2019/624262/EPRS_STU(2019)624262_EN.pdf

[20] Ray, S. (2019), Google, Amazon, Microsoft must disclose how they rank search results under new EU rules, Forbes, 7 December. Available at: https://www.forbes.com/sites/silad

Hattie Piemental (2021)[21] reporting on an online event organized by the Brookings Institute 'Should the government play a role in reducing algorithmic bias?', on 12 March highlights some of the processes that governments can adopt to quality check algorithms and minimize algorithmic bias. 'The process of checking algorithmic bias will be a cycle of identifying algorithmic requirements, researching how to build compliant algorithms, and then enforcing those requirements. Nascent compliance and risk assurance tools can emulate other industries: For compliance, governments can use verification, audit, certification, and accreditation tools; for risk, impact assessments, audits, and ongoing testing. Requirements will vary by industry and context; for instance, in environments with high uncertainty and variation, some disparate impact may be tolerable, but in others it must be rigorously avoided. Guidelines must combine principles with precise thresholds'.

Anticompetitive Behaviour and Its Regulation

One of the most long-standing reasons for the regulation of platforms is their perceived anticompetitive behaviours, a consequence of their power as incumbents across the search and sociality markets. As the four, country studies in this volume exemplify, the lack of competition, has led to all sorts of dependencies on key platforms by users, businesses, advertisers, and companies. In both the cases of the EU and India, data protectionism is a key reason for the advancement of platform regulation. In the EU, the space for European firms to be involved in search and sociality while in India the involvement of local players has, in the case of the latter, been shaped by the present government's 'swadeshi' (Made in India) rhetoric and preference for local monopolies such as Reliance over transnational platforms in the matter of the protection of locally produced data. It is in the United States, where these platforms are

ityaray/2020/12/07/google-amazon-microsoft-must-disclose-how-they-rank-search-results-under-new-eu-rules/?sh=6cf4bd9911f3

[21] Piemental, H. (2021), Should the government play a role in reducing algorithmic bias?, Brookings Institute, 12 March. Available at: https://www.brookings.edu/events/should-the-gov ernment-play-a-role-in-reducing-algorithmic-bias/

headquartered that their real dominance has led to bi-Partisan attempts to curb their monopoly power. While the Republicans and Democrats have differing reasons for wanting to regulate these monopolies, the fact that Google, for example, controls 92% of the global search market (88% in the United States), 85% of the Smartphone Operating System market (Android), 66% of the browser market and 28% of digital advertisements (70% in the United States)[22] and that it uses this power to incentivize companies such as Apple to favour Google search in Safari and to buy up companies such as YouTube, DoubleClick, and Android that now belong to its ecosystem, has led to the US Department of Justice filing an anti-trust ruling against Google in October 2020 that is comparable to similar filings against AT&T in 1974 and Microsoft in 1998. The Complaint highlighted Google's exclusionary ways of carrying out business and the detrimental impact of this on both competition and consumers:

- Entering into exclusivity agreements that forbid preinstallation of any competing search service.
- Entering into tying and other arrangements that force preinstallation of its search applications in prime locations on mobile devices and make them undeletable, regardless of consumer preference.
- Entering into long-term agreements with Apple that require Google to be the default—and de facto exclusive—general search engine on Apple's popular Safari browser and other Apple search tools.
- Generally using monopoly profits to buy preferential treatment for its search engine on devices, web browsers, and other search access points, creating a continuous and self-reinforcing cycle of monopolization.[23]

In spite of such clear evidence of anticompetitive behaviour there are those who have argued in the United States against such rulings in the context of fast-changing, novel market settings, and the potential error

[22] Ghaffary, S. & Molla, R. (2020), Why the US government is suing Google, Vox Recode, 20 October. Available at: https://www.vox.com/recode/21524710/google-antitrust-lawsuit-doj-search-trump-bill-barr

[23] Justice Department Sues Monopolist Google for Violating Antitrust Laws (2020), The US Department of Justice, Press Release, 20 October. Available at: https://www.justice.gov/opa/pr/justice-department-sues-monopolist-google-violating-antitrust-laws

costs of such decisions.[24] However, it would seem that the evidence of the untrammelled power of these platforms has led to bipartisan invoking of antitrust laws in the United States including The Sherman Act (1890) that deals with monopolies and conspiracies to monopolize, The Federal Trade Commission Act (1914) that bans unfair methods of competition and the Clayton Act (1914) that deals with the anticompetitive behaviour through mergers and interlocking directorates.[25] There are differences in the approaches taken to antitrust in the United States and EU. In the United States, market leadership is not the primary issue although greater efficiencies and cost savings for consumers are, whereas in the EU, there is a marked tendency to support potential and real local competitors. It has been argued that the protectionist stance taken by the EU needs to be seen in the context of the US$850 billion e-commerce sales within the EU by US companies, and the fact that EU platforms are finding it difficult to compete with their counterparts from the United States and also China. In the words of Kati Suominen (2020), 'Europe is using antitrust to clear space for its own companies in sectors it considers to be in Europe's comparative advantage, such as financial services, the Internet of Things (IoT), smart factories and smart homes, and healthcare. **Europeans have failed to seize** on the various technology waves that brought us smartphones, cloud computing, search, and social media, and they lack the kind of market-leading platforms that the United States and China have produced such as Amazon, Facebook, Twitter, Google, Alibaba, and WeChat. Germany's SAP, the Netherland's Adyen, and Sweden's Spotify have **barely 3 percent of the market capitalization** of major tech platforms compared to 68 percent held by U.S. companies'.[26]

EC antitrust policy is highlighted in Articles 101 (anticompetitive behaviour between two or more market operators) and 102 (abusive behaviour by companies in a monopoly position) of the Treaty of the Functioning of the European Union.[27] In June 2021, the EU opened a

---

[24] Manne, G. A. &Wright, J. D. (2011), Google and the limits of antitrust: The case against the antitrust case against Google (171–244), Harvard Journal of Law and Public Policy, 34 (1).

[25] The Antitrust Laws, Federal Trade Commission. Available at: https://www.ftc.gov/tips-adv ice/competition-guidance/guide-antitrust-laws/antitrust-laws

[26] Suominen, K. (2020), On the rise: Europe's competition policy challenges to technology companies, Centre for Strategic and International Studies, 26 October. Available at: https:// www.csis.org/analysis/rise-europes-competition-policy-challenges-technology-companies

[27] Antitrust, Competition policy. European Commission. Available at: https://ec.europa.eu/ competition-policy/antitrust_en

formal antitrust proceeding against Google for using its own display advertising technology services to skew the ad tech supply chain in its favour.[28] This proceeding is one against many levied by the EU against the anticompetitive cultures of Big Tech companies operating in the EU. The EU had previously imposed a record US$5 billion fine in 2018 on Google for stifling competition in the Android market, one of three antitrust rulings against Google.[29]

The Google Tax

The regulatory ideal that has unanimously been agreed to by most countries, including an initially reluctant the United States, is a tax regime and framework that explicitly enables the taxation of digital business-to-business (b2b) and business-to-customer (b2c) transactions. The theory and practice of this tax have been explored in the introductory chapter and the chapter on the Equalization Levy in India that is the Indian government's response to the Google Tax although its genesis as a global model for tax reform first proposed by the G20 can be traced back to the Organisation for European Development and Cooperation's (OECD) Base Erosion and Profit Shifting (BEPS) initiative that explicitly focussed on creating a framework to deal with a diverted profits tax (DPT) and tax avoidance by Big Tech companies based in the United States. This decade-long negotiation has led to the OECD's Inclusive Framework Agreement (IFA, 2021) that stipulates a 15% minimum corporate tax rate, a measure that will impact countries such as Ireland, which hitherto was a tax haven for Big Tech companies such as Google, Facebook, and Apple, three among 800 companies headquartered in Ireland and involved in substantive business in the EU but paying little taxes to the governments in the jurisdictions that they operate in.[30] Big Tech companies have habitually

---

[28] Antitrust, Commission opens investigation into possible anticompetitive conduct by Google in the online advertising technology sector, EU Press release, 22 June. Available at: https://ec.europa.eu/commission/presscorner/detail/en/ip_21_3143

[29] Google is appealing a $5 billion antitrust fine in the EU, The Associated Press, 27 September. Available at: https://www.npr.org/2021/09/27/1040889789/google-eu-android-appeal-antitrust

[30] IFA. (2021), Statement on a Two Pillar solution to address the tax challenges arising from the digitalization of the economy, OECD/Base Erosion and Profit Sharing Project, October, 8, 2021. Available at: https://www.oecd.org/tax/beps/statement-on-a-two-pillar-solution-to-addr ess-the-tax-challenges-arising-from-the-digitalisation-of-the-economy-october-2021.pdf

tried to avoid paying even minimal taxes. Google, for instance, used the cover of the 'double Irish' tax arrangement (abolished in 2015 but phased out only in 2020?) in 2019 to funnel US$75.4 billion of profits made by Google Ireland Holdings Unlimited Company via interim dividends and other payments to Bermuda where it is tax domiciled.[31] As cited earlier in this chapter, the OECD's IFA makes it mandatory for all companies to pay taxes in the jurisdictions where they trade their goods and services in even if they do not have a brick and mortar presence in those countries.[32] During the protracted BEPS negotiations, a number of countries including the United Kingdom in 2015 and Australia in 2016[33] along with Spain, Austria, France, India, and other countries created their own DPT instruments although in the light of the IFA, these unilateral measures, including the equalization levy in India will need to be withdrawn in 2023.[34]

While the regulatory moment seems to have truly arrived with the Chinese government too clamping down on its key platform Alibaba, levying a US$2.8 billion fine for abusing its market position and abandoning the proposed initial public offering (IPO) of its financial company Ant Financial,[35] it remains to be seen how regulation will be enforced and what these Big Tech companies will do to try and avoid or at least minimize the financial impact of regulation on their companies. It is already clear that some companies plan to pass on the financial burden incurred to their clients and consumers who will have to pay more for their access to these services. Advertisers on Google India, for example, will have to pay for the 2% surcharge incurred and included in

[31] Taylor, C. (2021), Google used 'double Irish' to shift $75.4 bn in profits out of Ireland, The Irish Times, 17 April. Available at: https://www.irishtimes.com/business/technology/google-used-double-irish-to-shift-75-4bn-in-profits-out-of-ireland-1.4540519

[32] Lyons, K. (2021), Ireland's status as tax haven for tech firms like Google, Facebook and Apple is ending, The Verge, 7 October. Available at: https://www.theverge.com/2021/10/7/22715229/ireland-status-tax-haven-google-facebook-apple

[33] Nielson, L. (2016–17), Diverted profits tax ('Google Tax'), Parliament of Australia, Available at: https://www.aph.gov.au/About_Parliament/Parliamentary_Departments/Parliamentary_Library/pubs/rp/BudgetReview201617/ProfitTax

[34] Choudhary, S. (2021), India awaits finality on taxing global MNCs before removing 'Google Tax', Business Standard, 11 October. Available at: https://www.business-standard.com/article/economy-policy/india-awaits-finality-on-taxing-global-mncs-before-removing-google-tax-121101000584_1.html

[35] Dans, E. (2021), Around the world, Governments are readying to regulate Big Tech, Forbes, 2 May. Available at: https://www.forbes.com/sites/enriquedans/2021/05/02/around-the-world-governments-are-readying-to-regulate-bigtech/?sh=2979df305935

the equalization levy.[36] This move by Google highlights the power of this platform to continue to maximize its profits at the expense of the companies that are dependent on them to sell goods and services. While Big Tech will try and use its economic and legal powers to bypass, stymie, and contest the regulation of platforms, both the State and civil society have mooted the ultimate regulatory action—the break-up of platforms. There are precedents such as the break-up of the US telecom monopoly AT&T in the 1980s. Nick Srnicek (2019) writing in *The Guardian* has made a case for cloud computing to be recategorized as an essential public good. 'we could decisively move away from Amazon, Google and Microsoft's control over cloud computing and imagine computing as a basic 21st-century utility. Initiatives such as the European Union's Open Science Cloud demonstrate one model for providing a publicly funded and operated cloud infrastructure. There's no reason why governments shouldn't provide citizens with free access to a public cloud that ensures privacy, security, energy efficiency and equal access for all. Our digital infrastructures should not be left in the hands of profit-seeking monopolies, but instead run democratically for the common good'.[37] There are however less drastic solutions. Given Google and Facebook's provisioning of what can be considered essential services that have become fundamental to citizens and their everyday practices of life, a case has been made to treat these companies as 'public utilities' and regulate these platforms deemed to be involved in the provisioning of public goods.

The three reasons for regulation highlighted in this introduction—the algorithmic power of platforms, anticompetitive behaviour, and non-payment of taxes are arguably three fundamental issues that have contributed to the regulatory moment. There are however other macro and micro issues related to platforms inclusive of privacy, their applications of AI, and their control over what we think and what we will think in the future that is also a cause for concern. Regulation needs to be seen as an important means by which the restructuring of the world based on

---

[36] Majumdar, R. (2021), Google to pass on Equalisation Levy to advertisers visible in India, Inc42, 21 July. Available at: https://inc42.com/buzz/google-to-pass-on-equalisation-levy-to-advertisers-visible-in-india/

[37] Srnicek, N. (2019), The only way to rein in big tech is to treat them as a public service, The Guardian, 23 April. Available at: https://www.theguardian.com/commentisfree/2019/apr/23/big-tech-google-facebook-unions-public-ownership

platforms can be shaped essentially by two entities—the State and civil society, who contribute in their own ways to the minimizing of 'risk' associated with the power of platforms and to making them, in that process, more transparent and accountable. While the private sector is involved in self-regulation, there is very little evidence to suggest that self-regulation has resulted in qualitative changes to content moderation and to voice equity. Arguably though, there is little compelling evidence to suggest that inter-governmental agencies such as the International Telecommunications Union or for that matter, multistakeholder organizations established on the lines of the Internet Consortium for Assigned Names and Numbers (ICANN) can be trusted to create regulatory frameworks for platforms. The governance of platforms remains an issue, and, at least for the moment, there is a lack of clarity related to who should be involved in governance and the best way forward.

2

Introduction to Platform Power

Over the last five years, sovereign countries including France, Germany, the United States, Australia, and India and regional entities such as the European Union (EU) have confronted and contested a variety of self-described 'platform exceptionalisms' including their descriptor as 'intermediaries' that is favoured by the platform industry, their expressed neutrality and contested their control over the global data economy. These confrontations have accentuated following the controversies related to the role played by these platforms in the gaming of politics, excluding competition, avoiding or minimalizing the payment of taxes and evidence of their emergence as sites of corporate, political, economic, and cultural power on a scale not previously enjoyed by corporate transnationals. Ireland's privacy watchdog has, for example, warned Facebook to desist from transferring data on its European users to the United States since it fell foul of EU's privacy laws (Scott 2020),[1] while the Australian government has informed content aggregators including Google and Facebook that they will be required to abide by a code of conduct and compensate traditional news media for their loss of revenues from hosting content produced by legacy media (see Hitch 2020).[2] The French government in 2020 has come to a settlement with Facebook for the payment of back taxes (106 million euros). Since 2017, the French government has negotiated a number of settlements with tech companies 'Amazon signed a €200 million check in 2018, Google agreed to pay €1 billion, Apple €500 million and Microsoft €350 million in 2019' (Braun 2020).[3] There are, in other words, a number of moves by nation-states

---

[1] Scott, M. (2020), Facebook to be forced to stop sending EU data to the USA, Politico, 9 September. Available at: https://www.politico.eu/article/facebook-privacy-data-us/

[2] Hitch, G. (2020), Google, Facebook will be forced to pay for news as part of new mandatory code of conduct to support traditional media, ABC, 31 July. Available at: https://www.abc.net.au/news/2020-07-31/draft-mandatory-code-conduct-facebook-google-pay-for-news/12510776

[3] Braun, E. (2020), Facebook to pay €106m in back taxes in France, Politico, 24 August. Available at: https://www.politico.eu/article/facebook-to-pay-e106-million-in-back-taxes-in-france/

Platform Regulation. Pradip Ninan Thomas, Oxford University Press. © Pradip Ninan Thomas 2023.
DOI: 10.1093/oso/9780192887962.003.0002

and regional blocs such as the EU to limit the overwhelming power of a handful of data controllers and to lay down rules that would strengthen both competition and protect the rights of individuals over their personal data. In Australia, following the Digital Platforms Inquiry in 2019 (see Chapter 7), there has been an attempt to create a Digital Platforms Regulators Forum (DP-REG) consisting of four stakeholders including the Australian Communications and Media Authority (ACMA), the Australian Competitions and Consumer Council (ACCC), the Office of the Australian Information Commissioner (OAIC), and the Office of the eSafety Commissioner to share information on a variety of data issues including privacy and competition and to strengthen ex ante and smart pre-emptive actions in a rapidly changing environment in which Big Tech share many advantages including their power over multiple resources— economic, human, technological, social.[4] This initiative is similar to the United Kingdom's Digital Regulation Co-operation Forum and its specific focus on Big Tech—'large digital platforms, including internet search engines, digital content aggregators, social media services, private messaging services, media referral services and electronic marketplaces' (Burton 2022).[5]

Regulation as Risk Management

Regulation, in other words, has come to the fore as a public discourse aided to a large extent by real and imagined fears related to information monopolies, the anti-competitive behaviours of Big Tech companies, their political influence, and non-payment of taxes in the jurisdictions that they operate in among other concerns. Their role in the influencing of electoral outcomes in the United States in 2016 and in the United Kingdom with respect to Brexit and the trail of evidence involving external actors such as Russia and firms such as Cambridge Analytica

---

[4] Coade, M. (2022), New form set up for Australia's digital regulators, The Mandarin, 18 March. Available at: https://www.themandarin.com.au/184487-new-forum-set-up-for-austral ias-digital-platform-regulators/

[5] Burton, T. (2022), Regulators join to form defence ring against digital platforms, The Financial Review, 13 March. Available at: https://www.afr.com/politics/federal/regulators-join-up-to-form-defence-ring-against-digital-platforms-20220313-p5a466

have highlighted the networked role played by platforms in global politics. The expansive and 'deep' power of Big Tech companies is arguably a reflection of market failures—a prime reason for regulation although in addition, there is an argument that these Big Tech companies function like public utilities hence the need to regulate them in the interests of the public good. It has become amply clear that the new regimes of data-driven capital accumulation have led to a social and economic crisis that demands renewed attempts to regulate this sector and create a level playing field for firms, consumers, and states (Beiling, Jager, & Ryner 2016).[6] Regional bodies such as the EU recognize the need for a risk management approach in data-driven societies and expressions such as the General Data Protection Regulation (GDPR) are focussed on regulation as standard setting, monitoring, behaviour control. They recognize the need for algorithmic accountability in a context characterized by Big Data 'hyper-nudgings' of consumers within architectures based on automated decision-making and digital decision-guidance processes (see Yeung 2017:123).[7] Gellert (2020:143),[8] in a volume entitled a *Risk-based Approach to Data Protection* observes that it not at all surprising 'that all the regulatory issues plaguing data protection are addressed in terms of accountability. This just confirms that accountability as the data protection-oriented embodiment of responsibility is indeed the plinth and keystone of any collaborative model of regulation'. Baldwin, Cave, and Lodge (2015:2–3)[9] highlight five questions related to effective regulation.

- Is the action or regime supported by legislative authority?
- Is there an appropriate scheme of accountability?
- Are procedures fair, accessible, and open?
- Is the regulator acting with sufficient expertise?
- Is the action or regime efficient?

---

[6] Beiling, H-J., Jager, J., & Ryner, M (2016), Regulation theory and the political economy of the European Union (53–69), Journal of Common Market Studies, 54 (1).

[7] Yeung, K. (2017), 'Hypernudge': Big data as a mode of regulation by design (118–136), Information, Communication & Society, 20 (1).

[8] Gellert, R. (2020), The Risk-based Approach to Data Protection, Oxford Scholarship Online, Oxford University Press.

[9] Baldwin, R., Cave, M., & Lodge, M. (2015), Understanding Regulation: Theory, Strategy and Practice, Oxford Scholarships Online, Oxford University Press.

A fundamental characteristic of social media and search platforms is their scale of operations and the fact that every connected user is potentially also a customer. This customer is not a consumer of one-way communications but is involved in multiple transactions that are value laden in environments such as the Facebook page and Google search that functions both explicitly and implicitly as sites for value creation. However, this value creation has multiple dimensions. While these are sites where buying and selling take place, they are also sites where cultural practices and branding are congealed, hidden, implicit, and where user data is consistently used as the raw material by these firms to better understand the transactional consumer—his/her sentiments, motivations, preferences that can form the basis for a more targeted and integrated projections of user behaviour vis a vis and platforms. This relatively new way of value creation is conditioned by new technological architectures and the possibilities that code and algorithms offer to shape the organization of society, politics, and the economy and the laws that govern this shaping. Paul Timmers (2019:18)[10] in an article on digital sovereignty has suggested the need for a change from 'code is law', the title of a classic book by Lawrence Lessig (1999), to 'law is code' and one in which the 'rules we want to have as a society start to condition technology architectures'. The question following on from this suggestion is whether governments are willing or able to reverse code is law to law is code in contexts in which information/surveillance capitalism's influence is deep and wide across all productive practices and sectors. In other words, if the platform economy is so deeply entrenched—what possibilities are there for lawmakers in different jurisdictions to make sovereign judgements linked to the regulation of the major companies involved in the platform economy? To make matters more complicated, there is typically an ambivalence in the attitude of individual nation states to platforms since political parties benefit from their use and abuse of social media. President Trump highlights this ambivalence—while he is an avid user of social media and resists any attempt to downsize these industries, he has, in the past, openly threatened Twitter for censuring his posts and has threatened trade sanctions against

[10] Timmers, P. (2019), Challenged by 'digital sovereignty' (1–20), Journal of Internet Law, 23 (6).

countries that have either levied or are contemplating the levying of a Google Tax.

So what has caught the attention of governments and regulators includes the fact that these companies and their products and services have begun to impact on the old economy—as, for example, legacy journalism that has been wounded by the migration of advertising revenues to these new platforms, stifling competition, and heralding an era in which algorithms have begun to a fundamental role in how information is being circulated and consumed. The rules have changed dramatically and the media industries in particular have had to face up with the consequences of disruption resulting in either the closure or digitization of the rural press in Australia, massive job losses in media industries in India and cinema releases over Netflix at the expense of multiplexes and traditional cinema halls, all trends that have accelerated in the COVID-19 environment. Moreover, there is mounting evidence that companies such as Amazon have begun to consolidate their power through investing in technological, contractual, and commercial architectures that offer parallel means to regulate third-party commercial users of this platform. Teachout and Sussman (2020)[11] writing in *The American Prospect*, highlight one of Amazon's new patents. 'The patent, for a form of blockchain ledgering technology, will allow Amazon to oversee the collection of an unprecedented amount of data about the business operations of its sellers, including their *entire supply chain*. In essence, the patent fulfils Amazon's plans to create a private regulatory regime, where it uses proprietary information to create a "certification" bureaucracy: a private, for-profit alternative to the Food and Drug Administration, the Environmental Protection Agency, and the Federal Trade Commission. Unlike governmental agencies, however, it will have no public oversight, and can use its certifying power to squeeze sellers and consolidate control'. There also is increased global concern with the fact that many of these companies have continued to evade or minimize paying taxes in the jurisdictions that they operate in along with what has been a dominant concern— issues with data privacy and surveillance that have been long-standing concerns. While there certainly have been issues with the deleterious

---

[11] Teachout, Z. & Sussman, S (2020), Amazon's private government, The American Prospect, 18 June. Available at: https://prospect.org/power/amazons-private-government/

effects of monopolies such as Standard Oil and AT&T in the past, the scale, depth, and embrace of these new companies have been unprecedented since their impact is felt across the world by ordinary people and across all productive sectors of the global economy. The fact that these companies feed on and profit from the energies, sentiments, behaviours, interests, and curiosities of their clients in the public and private sectors and civil society suggest that regulation will remain a challenge for some time yet precisely because governments have not previously had to deal with the ramifications of such comprehensive and deeply entrenched global companies.

Regulation Is in Process

Despite all the disruptions that have been caused over the ensuing decade, it is only over the last few years that governments have begun to come to terms with the need to provide rule-based frameworks for the operations of this handful of globally significant companies. The approach followed by governments has been inconsistent precisely because of the fact that social media platforms in particular have been used by political parties to their electoral advantage, as has been the case of Facebook in India and Twitter in the United States. Economies of scale have been achieved by political parties that have used these platforms to reach out to citizens before others within unregulated environments. With both social media platforms and political parties benefiting from these regulation-lite environments, it has been difficult to get bi-partisan support for any pronounced regulation. So, for the most part, in many countries, with the exception of the EU and countries within the EU such as France, Germany, and Spain, regulation has been ad hoc, perfunctory, and incomplete.

Alex Rochefort (2020),[12] in an article on the regulation of social media platforms, highlights the fact that all policymaking is constrained by four considerations: the effectiveness of policy goals, the administrative difficulty in implementing policy, the costs of implementing policy, and the

---

[12] Rochefort, A. (2020), Regulating social media platforms: A comparative policy analysis (225–260), Communication Law & Policy, 25 (2)

political acceptability of any given policy (234). Arguably, these criteria, drawn from a rational approach to policymaking, has to also take into account the fact that the very practice of politics today is dependent on and shaped by Facebook, Twitter, Instagram, and YouTube. It is this enmeshed and closely integrated spiral of interests that have hitherto remained an obstacle to the making of any effective regulation of the platform economy. Rational theories of policymaking simply cannot be used to make sense of contemporary regulations of platforms because such ideal type constructions are incapable of accounting for the many 'irrational' factors that condition policymaking and/or the lack of it. Lobbying, special tax exemptions, investments in corporate social responsibility projects, and clear political uses of social media have enabled these companies to resist significant regulation. Moreover, any regulatory policy specifically targeted at these tech companies outside of the United States has been contested by US state bodies such as the US Trade Representative (USTR) and Section 301 provisions, the US Agency for Global Media (USAGM) support for internet freedoms and by the example of the United States withdrawing from negotiations on the Global Digital Tax during the Trump era (although the Biden administration has reversed this decision) that is supported by 140 member countries of the Organisation for Economic Cooperation and Development (OECD), ostensibly because of the more pressing need to fight COVID-19 (see Leonard & Davison 2020).[13]

So it would seem that a global move to regulate big media platforms is in process although it will take time for these regulatory frameworks and laws to become normalized. In Australia, the Australian Consumer and Competition Commissioner's (ACCC) 2019 recommendations to regulate platform media have been taken up the government although Facebook and Google are resisting this attempt at regulation. In India, Facebook's intimate relationship with the BJP government including evidence of conflicts of interest between Facebook employees embedded in the incumbent government's electoral strategy in 2014 has been reported widely. The reports have focussed on the role played by Facebook's public

---

[13] Leonard, J. & Davison L. (2020), U.S. has pulled out of global digital tax talks, Lighthizer says, Bloomburg/Quint, 18 June. Available at: https://www.bloombergquint.com/global-economics/u-s-pulls-out-of-global-digital-tax-talks-lighthizer-says

policy head Ankhi Das, and her rather explicit support for PM Modi, disparagement of the opposition Congress party, unwillingness to take down hate speech and anti-Muslim comments (The Wire Staff 2020).[14]

Regulation, the State, and Civil Society

While the State does have a significant role to play in Platform governance, it will be interesting to see if this will be negotiated through multilateral as opposed to multistakeholder processes. The latter has been the preferred option in the governance of the internet although there is little evidence that 'decision-making' has inherently and consistently included civil society. There have been robust discussions on both internet and platform governance on the JustNetCoalition forum by a variety of informed activists located in various parts of the world. To quote Maclean (2004:18–19),[15] 'The most powerful actors are able to exercise a significant degree of policy and regulatory control "from the bottom up" by pursuing national and/or regional interests across a wide range of fora, while the most powerful private actors are able to exercise an equally significant degree of market control by coordinating their activities through private fora, or through the exercise of raw market power. But what is often missing are opportunities for the less powerful to be engaged in discussion of global governance issues, to participate in decision-making processes, to understand the consequences of these decisions, and to adapt their policies, regulations and practices accordingly'.

Regulating Platforms: Contested Views

While current moves to regulate platforms have been a direct consequence of the perceived and real influence of a handful of corporations in the shaping of the global economy, politics, and culture, there have

[14] Wire Staff (2020), News report says Facebook's Ankhi Das supported Modi, hoped for BJP's victory, The Wire, 31 August. Available at: https://thewire.in/tech/facebook-ankhi-das-modi-bjp-support-wsj-new-report

[15] Mclean, D. (2004), Herding Schrodinger's Cats: Some conceptual tools for thinking about internet governance (1–26), Background Paper for the ITU Workshop on Internet Governance Geneva, 26–27 February 2004.

been significant voices against regulation—from positions expressed by Judge Frank Easterbrook (1996) to those who have argued that the law is inherently incapable of regulating fast-changing, data-driven technologies that are based on exponential innovation and that it will result in the stifling of creativity. Easterbrook (215–216)[16] famously declared that 'Error in legislation is common, and never more so than when the technology is galloping forward. Let us not struggle to match an imperfect legal system to an evolving world that we understand poorly. Let us instead do what is essential to permit the participants in this evolving world to make their own decisions'. He argued that just as a course on the law of the horse would be doomed to fail because horses and their behaviours can be explored from myriad legal perspectives, so would attempt to regulate cyberspace. Easterbrook's position was challenged by the Stanford lawyer Lawrence Lessig (1999)[17] who contended that behaviours in cyberspace were constrained by laws such as copyright, by norms, markets, and code precisely because this space was an ambit of value-laden human activity. He argued that 'As the net grows, as its regulatory power increases, as its power as a source of norms becomes established, the values of real space sovereigns lose. In many cases, that is a good thing. But there is no reason to believe that it will be a good thing generally. There is nothing to guarantee that the regime of code will be a liberal regime; and little reason to expect an invisible hand of codewriters to push it in that way. Indeed, to the extent that code writers respond to the wishes of commerce, a power to control may well be the tilt that this code begins to take' (548). Shelanski (2013:1668)[18] has disputed the use of anti-trust that has evolved from an understanding of competition issues in the analogue era to deal with the digital era and in favour of understanding the 'characteristics that differentiate competition on the Internet' to that expressed in conventional anti-trust laws. Shelanski, however, veers towards a rather uncritical support for this consumer-data driven innovation-based model and its whole market dominance approach, is against overenforcement while having little to

---

[16] Easterbrook, F. H. (1996), Cyberspace and the Law of the Horse (207–216), The University of Chicago Legal Forum, 207.

[17] Lessig, L. (1999), The Law of the Horse: What cyberlaw might teach (501–549), Harvard Law Review, 113 (2).

[18] Shelanski, H. (2013), Information, innovation and competition policy for the internet (1663–1705), University of Pennsylvania Law Review, 161 (6).

say about a model that is inherently built on the mining and maximiza-tion of user generated data while minimalizing privacy concerns. M Napoli and A Napoli (2019)[19] have held the view that social media could adopt the self-regulatory, audience measurement-based model from legacy media building on their attempts to regulate hate speech, polit-ical interference, etc., although arguably, it is precisely the unwillingness shown by these companies to any type of regulation that has prompted government moves to regulate this industry. It is also clear that that these platforms are more sophisticated than legacy media given the fact that their data-driven, algorithmic influence is deep, multisectoral, and multidimensional, embracing and deriving surplus value from so-cial, human interactions, and affect. Hence the need for regulation that brings these corporate giants within acceptable norms and rules sup-portive of the public good. Facebook and other data aggregators are involved, as Schwarz (2019:135)[20] has argued, in no less than the govern-ance of digital capitalism 'The governance toolkit used by Facebook and other digital corporations is highly eclectic, bringing together sovereign legislation and design choices, seduction and disciplinary surveillance, constitutionalism and A/B testing, algorithmic governance and prole-tarianized administration of justice'. Platform governance and their al-gorithmic 'black boxes' in other words, need to be brought within the ambit of global governance mechanisms that monitor the need for com-petition, a level playing field, private profit, and the public good. It is the case that contemporary platforms function as private licensing regimes. McKee (2018:38)[21] has argued that such regimes may end up reprodu-cing 'the pathologies associated with publically mandated licensing. They may place providers in a precarious position, and they are unlikely to operate in the public interest'. Van Dijk (2013:38–39)[22] has com-mented on the governance functions in end-user license agreements and

[19] Napoli, P. M. & Napoli, A. (2019), What social media platforms can learn from audience measurement: Lessons in the self-regulation of 'black boxes', First Monday. Available at: https://doi.org/10.5210/fm.v24i12.10124

[20] Schwarz, O. (2019), Facebook rules: Structures of governance in digital capitalism and the control of generalized social capital (117–141), Theory, Culture & Society, 36 (4).

[21] McKee, D. (2018), Peer platform markets and licensing regimes (17–53), in McKee, D., Makela, F., & Scassa, T. (Eds.), Law and the 'Sharing Economy': Regulating Online Market Platforms, University of Ottawa Press, Ottawa.

[22] Van Dijk, J. (2013), The Culture of Connectivity: A Critical History of Social Media, Oxford University Press, New York

terms of service that are often unread by users and the changes in governance that scarcely involve any consultation with its users. Terms of service as van Dijk has pointed out includes the right of platform entities to use and monetize data provided by its users. All users are locked when they log into a site and they forfeit the right to control access to their metadata.

Regulation and the Public Good

The role of the State in platform governance is an evolving one and ranges from a digital welfare state in which access, availability, dialogical rights, and privacy are core principles in the administration of welfare such as is the case in some Nordic countries such as Finland (see Ala-Fossi et al. 2019:4)[23] to the situation prevailing in many parts of the world where state involvement in platform governance is episodic and inconsistent as the case in India where the relationship between Facebook in particular and the government is mutually beneficial. Andrejevich (2013)[24] has argued for a more expansive public interest, data access eco-system involving the public library system, public broadcasting, and other public utilities as an alternative approach to ensuring data access for all, thus reinforcing data abundance rather than data scarcity. Arguably, a regulated platform economy informed by public interest principles such as neutrality and universal access should not be discounted because of the prevailing architectures of platform data flows that are global and extensive. Srnicek (2017:47)[25] highlights five types of platforms—advertising platforms, cloud platforms, industrial platforms, product platforms, and lean platforms that are key to digital capitalism. Srnicek too is of the view that in contrast to the exertion of state control over platforms that he finds 'unimaginative and minimal' (128), that state efforts and investment must be made in the creation of parallel platforms that support the

[23] Ala-Fossi, M., Alen-Savikko, A., Hilden, J., Horowitz, M A., Jaasaari, J., Karppinen, K., Lehtisaari, K., & Nieminem, H. (2019), Operationalising communication rights: The case of a 'digital welfare state' (1–17), Internet Policy Review, 8(1).
[24] Andrejevich, M. (2013), Public service media utilities: Rethinking search engines and social networking as public goods (123–132), Media International Australia, 146 (1).
[25] Srnicek, N. (2017), Platform Capitalism, Polity, Cambridge/Malden.

public sphere, are controlled by citizens and that outside of the panoptic gaze of mainstream platforms (129).

My own position that is probably not at all surprising to readers of my works is that an interventionist approach is absolutely critical given the extraordinary evidence of the role played by algorithmic capitalism in expanding surveillance, gaming electoral outcomes, and the monetization of affect. There is little evidence that in-house based, private, regulation initiatives have advanced the public good. On the contrary, the platform economy has been profoundly disruptive to the wage–labour relationship since the gig economy is built on precarity as a natural set of circumstances that is critical to competition and innovation. This sleight of hand used by those who advocate non-intervention in platform regulation is ultimately self-defeating since the self-reproduction of labour has been essential to capitalist growth, a key aspect of the crisis of capitalism and absolutely vital to its future growth. A working class unable to reproduce themselves seems a rather catastrophic condition for capitalism that has thrived on the provisioning of minimum, subsistence support for working classes who have been able to labour for capital. It would seem that global capitalism based on the gig economy will be met by a 'double movement', a phrase used by Polyani (2001)[26] to describe public recognition of the failure of the market to enable a just distribution of goods and services that can also be applied to the just availability of digital goods and services. Polyani had highlighted the Poor Law Reform Act of 1834 that led to the establishment of an organized labour market and in 19th century England, a raft of measures supportive of child protection, workplace safety, and public libraries that attempted to provide a social net for those who had not been served by a self-regulating market economy. In Polyani's words (2001:87), 'factory laws and social legislation, and a political and industrial working-class movement sprang into being. It was in this attempt to stave off the entirely new dangers of the market mechanism that protective action conflicted fatally with the self-regulation of the system. It is no exaggeration to say that the social history of the nineteenth century was determined by the logic of the market

---

[26] Polyani, K. (2001), The Great Transformation: The Political and Economic Origins of Our Times, Beacon Press, Boston.

system proper after it was released by the Poor Law Reform Act of 1834'
(see also Maertens 2008).[27] The double movement that the welfare state
in the United Kingdom was involved in providing enabled a State-based
correction of the inability of a self-regulating market to provide essential
services to all citizens—access to food, a roof above one's head, employment, and self-reproduction. Just as Polyani insisted that the economy
be embedded in the social, one can argue that the platform economy too
needs a regulatory framework that enables platform neutrality and possibilities for it to be socialized (see Cioffi et al.'s 2022 article on regulation
as a 'double movement').[28] While I do acknowledge the complex nature of
issues related to platform regulation, I am not in favour of a supposedly
objective, 'let us hear all sides of the argument' type approach to dealing
with platforms precisely because of the deep power of a handful of platforms that now control behaviours, sentiments, and the everyday affective lives of ordinary people throughout the world. They are no longer
in an innovations mode that may have warranted an approach based on
taking seriously all sides of the argument. However, these companies have
been found wanting on any number of occasions and they have taken a
rather cavalier attitude towards responding to many concerns that have
been raised by regulators around the world. Regulators and policymakers
around the world in numerous jurisdictions have raised issues including
censorship and hate speech, anti-competitive behaviour, neutrality, tax
avoidance, the gaming of elections, data-privacy, and while there have
been a few financial settlements in the EU in particular and attempts to
curb hate speech, for the most, such moves have been episodic, limited to
specific jurisdictions while they have ignored the systematic abuse that is
inherent to algorithmic control. The key role that they play in the struc-
turing of data capitalism and the global economy is unprecedented hence
the need for regulation that is supportive of a level playing field and the
larger public good.

---

[27] Maertens, E. (2008), Polyani's double movement: A critical reappraisal (129–153), Social
Thought & Research, 29 (Globalization).
[28] Cioffi, J. W., Kenney, M. F., & Zysman, J. (2022), Platform power and regulatory politics
(1–17), New Political Economy, https://doi.org/10.1080/13563467.2022.2027355

The following two examples of platform regulation highlight the nature of investments in regulation in two jurisdictions, namely the EU and in the state of California in the United States.

The General Data Protection Regulation (GDPR)

The EU's GDPR that came into force in May 2018 is the successor to the EU Data Protection Directive (1995), and recognizes privacy and the protection of personal data as fundamental rights of all EU citizens.[29] As a regulation rather than a directive, it provides a single legal basis for data protection applicable to all EU nations including those who are not part of the Euro zone. The GDPR to date is the world's most advanced data protection instrument that also provides the basis for data transfers within a single digital market. These rules also apply to personal data transfers to countries outside of the EU. Two controversial principles that are at the very core of the GDPR are data portability and the right to be forgotten. Data portability refers to citizens' rights to receive one's personal data in a commonly used, machine-readable format, without any hindrance from one service provider to another, and Article 17, the right to the complete erasure of all personal data—the right to be forgotten (Article 17) 'The data subject shall have the right to obtain from the controller the erasure of personal data concerning him or her without undue delay and the controller shall have the obligation to erase personal data without undue delay') (Art.17:GDPR). Data protection by design and default is the basis for a normative framework that data processors and controllers in the EU are expected to incorporate into their overall business strategy in the EU. The penalties for data mismanagement are high—a fine that can be up to 4% of a firm's global revenues or Euros 20 million, whichever is higher. The GDPR, as expected, has riled the United States in particular given the ramifications for US companies such as Amazon, Google, Facebook, and others that thrive on their ability to transfer, mine, monetize and trade in personal and non-personal data. With the exception of the California Consumer Privacy Act (2020), which does provide data providers control over how data is used by data companies, US privacy laws at both federal

---

[29] Art 17:GDPR, Intersoft Consulting, Available at: https://gdpr-info.eu/art-17-gdpr/

and state levels are weak. EU–US data flows were until 2013, based on the Safe Harbour scheme—voluntary self-certification under the aegis of the US Department of Commerce and Federal Trading Commission. Three thousand two hundred companies had been certified under this scheme. However, the Snowdon revelations in 2013 that implicated the National Security Agency in data surveillance with the willing support of all the major US-based data companies resulted in serious misgivings at the EU's Court of Justice of the European Union (CJEU) (see Tzanou 2017).[30] The GDPR's regulatory framework is also at odds with the trade in services framework provided by the General Agreement on Trade in Services (GATS, 1994) and in which privacy is secondary to the primacy of unfettered, global trade (see Yakovleva & Irion 2020).[31] The GDPR also, at a fundamental level, has questioned the agenda-setting role of algorithms and machine learning and forced big data companies to pay more attention to the need to factor in fairness and the public good in the creation of algorithms—an issue that is also applicable to algorithms being used in the public sector to validate or for that matter proscribe welfare recipients. As cited earlier algorithms were used in the United Kingdom in 2020, to game the ranking of school leavers resulting in the grades of students from lower-income post-code areas being literally downgraded. As Maxwell and Tomlinson (2020),[32] writing in *The Prospect* on the A level grading fiasco, point out that 'there must be rigorous, expert and independent scrutiny of government algorithms before they are deployed. Transparency alone will not ensure that algorithms are accurate, fair, and lawful. This is particularly so where, as here and in many cases, an algorithm is technically complex and difficult for laypeople to examine, and where, unlike here, an algorithm affects only a discrete minority with limited institutional support. There must also be robust legal and policy structures to support this scrutiny'.

---

[30] Tzanou, M. (2017), European Union regulation of transatlantic data transfers and online surveillance (545–565), Human Rights Law Review, 17.

[31] Yakovleva, S. & Irion, K. (2020), Pitching trade against privacy: Reconciling EU governance of personal data flows with external trade (201–221), International Data Privacy Law, 10 (3).

[32] Maxwell, J. & Tomlinson, J. (2020), What the A-level debacle teaches us about algorithms and government, The Prospect, 26 August. Available at: https://www.prospectmagazine.co.uk/science-and-technology/a-level-students-exams-algorithm-ofqual-government-gavin-williamson

Assembly Bill 5 (AB5): Platform Labour in the Gig Economy

Another recent attempt at creating a level playing field is Assembly Bill 5 (AB5)[33] passed by the government in California and that specifically reclassifies gig workers as employees who are eligible for a minimum wage, overtime, and paid sick leave. While this may affect workers in the ridesharing economy such as Uber and Lyft differently, those who work in this sector and are part of the precariat will benefit since the major aim of this bill is to ensure workers' rights across sectors. Section 1 of the Bill highlights the specific issue that workers face in the gig economy:

(b) In its decision, the Court cited the harm to misclassified workers who lose significant workplace protections, the unfairness to employers who must compete with companies that misclassify, and the loss to the state of needed revenue from companies that use misclassification to avoid obligations such as payment of payroll taxes, payment of premiums for workers' compensation, Social Security, unemployment, and disability insurance.

(c) The misclassification of workers as independent contractors has been a significant factor in the erosion of the middle class and the rise in income inequality.

(d) It is the intent of the Legislature in enacting this act to include provisions that would codify the decision of the California Supreme Court in Dynamex and would clarify the decision's application in state law.

(e) It is also the intent of the Legislature in enacting this act to ensure workers who are currently exploited by being misclassified as independent contractors instead of recognized as employees have the basic rights and protections they deserve under the law, including a minimum wage, workers' compensation if they are injured on the job, unemployment insurance, paid sick leave, and paid family leave. By codifying the California Supreme Court's landmark, unanimous Dynamex decision, this act restores these important protections to potentially several

---

[33] AB-5 Worker status: employees and independent contractors (2019–2020), California Legislative Information, Assembly Bill No. 5, Chapter 296. Available at: https://leginfo.legislature.ca.gov/faces/billNavClient.xhtml?bill_id=201920200AB5

million workers who have been denied these basic workplace rights that all employees are entitled to under the law (Assembly Bill No. 5).

This bill was a response to moves by companies to change the status of the workforce from employees to contractors, often illegally to increase profits (see Roosevelt 2020).[34] Wage inequality in the gig economy has been a key concern in California and other jurisdictions in the United States, and the disruption to traditional employment caused by platforms has been an issue that has long been of public concern. With apps setting the price, customer credit cards being charged by the company who then pay the worker, the company controlling wages and price while the worker retains control over hours of work (Sprincin 2019)[35]—there was bound to be abuse and in particular the steady lowering of wages that has been experienced by Uber drivers throughout the world. The Bill is focussed squarely on those involved in the new gig economy and not those in the music industry, for example, who invariably work as contractors although the Bill has provided an opportunity to the industry to revisit the issue of a living wage for musicians and the right to appeal to the State for compensation if they were being paid less than the minimum wage (Aswad 2020).[36] As Schor et al. (2020)[37] have argued

> If platforms become less tolerant of supplemental earners and the proportion of dependent workers grows over time, satisfaction, hourly wages, and autonomy will decline. Conversely, the availability of alternative options in the larger labor market will regulate this pressure. If jobs are plentiful elsewhere, platforms will be forced to improve conditions. The converse is true for slack labor markets. If substantial

---

[34] Roosevelt, M. (2020), New California Labor Law AB5 is already changing how businesses treat workers, Los Angeles Times, 14 February. Available at: https://www.latimes.com/business/story/2020-02-14/la-fi-california-independent-contractor-small-business-ab5

[35] Sprincin, P. (2018), For hire backfire, City Journal, 13 September. Available at: https://www.city-journal.org/california-assembly-bill-5

[36] Aswad, J. (2020), Musicians to be exempt from California 'Gig Economy' Assembly Bill 5, Variety, 17 April. Available at: https://variety.com/2020/music/news/california-gig-economy-assembly-bill-5-ab5-musicians-1234583320/

[37] Schor, J. B., Attwood-Charles, W., Cansoy, M., Ladegaard, I., & Wengronowitz, R. (2020), Dependence and precarity in the platform economy, Theory & Society, Available at: https://doi.org/10.1007/s11186-020-09408-y

diversity does persist it is likely there will be more public attention to the fact that platforms are free-riding on conventional employers who offer full-time work and benefits and platforms may be forced to shoulder costs they are now externalizing.

However, the future of regular employment in the context of COVID is bound to remain dire for some time yet given the cross-sectoral toll that this pandemic has taken. India, which was on an upward growth trajectory until 2017, followed by a decline has, in the wake of an ill-thought-out lockdown that negatively affected migrant labour in particular is now facing unemployment in the region of 35%. In an economic context characterized by underemployment across all sectors except in agriculture, it would seem more likely than not that reforms to the gig economy will be on hold since there is a need for employment, any type of employment. Apart from ongoing reforms, any attempts to introduce any substantive new reforms to this sector could well be placed on hold across the world as countries try and deal with what has been the catastrophic contraction of all economies. COVID has of course provided the perfect opportunity for these platforms to expand on their business logic that is based on virtual solutions. The demand for e-commerce and sociality has expanded greatly in the context of a lockdown where people have been forced to live contained lives. In other words, the global lockdown has provided the perfect environment for these platforms to consolidate their interests and business strategies and this has led to these platforms becoming more dominant position today than they were in any other period during their relatively short history. Arguably though, this situation also provides an opportunity for governments to strengthen the terms of employment across multiple sectors since this would be a far better option than dealing with the consequences of unemployment based on perpetual precarity.

Platform-user relationships are typically based on asymmetries of information that are powered by machine learning algorithms. The role of these algorithms is to discriminate, to rank, hierarchise, order, privilege information, and by default the consumers of information. This can result in algorithmic oppression that is typically difficult to police because of the opaque nature of how algorithms are designed. Mueller (2020:178)[38] in an article on the algorithmic harms to Uber drivers has highlighted 'Evidence from multiple studies and experiments suggests that platforms that incorporate a reputational or rating component will often reflect the racial biases of their service providers: for instance, a study on Airbnb revealed that guests with African-American sounding names are sixteen percent less likely to have their reservation requests accepted, while another study focusing on Uber showed that female and African-American passengers suffered from various forms of discrimination by drivers, including longer wait times, increased cancellations, and, particularly in the case of female riders, more circuitous routes taken by drivers'.

There is from the perspective of the State, an equally compelling need to regulate the platform economy—the fact that data aggregators who are often chary of the need for privacy of user generated data, contribute to and heighten the nature of risk in a risk society.

---

[38] Mueller, Z. Algorithmic harms of workers in the platform economy: The case of Uber (167–210), Columbia Journal of Law and Social Problems, 53 (2).

3

The EU, Platform Regulation, and the 'Right to Be Forgotten'

Writing this chapter in the last days of the Trump administration in the United States and the suspension of his accounts on Twitter and Facebook, it is important to not forget the fact that such actions were preceded by many years of inaction by these large data controllers on the peddling of hate speech online. Every transaction online including hate speech is potentially an opportunity for the monetization of data and the many communities of practice associated with right-wing extremism have been sources of income for data controllers such as Amazon, Google, Facebook, and other Big Tech firms. All data, genomic, medical, private, public, past, present, and future, has become essential to the basis for and the expansion of what one might call 'curative capitalism'. It is this expansion and the control and ownership of this data in the hands of a handful of global companies that have motivated countries to invest seriously in the regulation of these companies. For the European Union, however, it is not just the power and reach of these companies that are issues to be dealt with but is centrally a reckoning with how to manage 'data' as currency, natural resource, and as the basis for identity in the context of trade and the protection of privacy within the Union and outside of it.

The General Data Protection Regulation (GDPR, 2018)[1] that is considered the gold standard in global data regulation today is the culmination of many decades of negotiation among member states to create a level playing field for data aggregators and data controllers of European and non-European origin who have substantial economic interests in data generated within the EU. The *Convention for the Protection of Individuals with Regard to Automatic Processing of Personal Data* (1981)

---

[1] General Data Protection Regulation. Available at: https://gdpr-info.eu/

Platform Regulation. Pradip Ninan Thomas, Oxford University Press. © Pradip Ninan Thomas 2023.
DOI: 10.1093/oso/9780192887962.003.0003

was a precursor to the EU Data Privacy Directive (1995). This was followed by the issuing of a number of regulations related to data processing within the single market, privacy, and retention of data. In 2010, the Data Protection Division of the EC's Directorate General for Justice was established with the mandate to harmonize policies within the EU (Valetini 2017).[2] As Newman (2008:93–94)[3] has observed, there was resistance from member states to a pan-EU data governance regime until the early 1990s and there were five full members of the Union inclusive of Greece, Spain, Portugal, and Belgium who were forced to create national regulatory bodies by 1998 as per the terms of the directive.

'First, all member states adopted comprehensive national legislation covering the public and private sectors. Second, the directive required that a national, independent control institution be created with the power to implement and enforce data privacy rules. Third, non-EU countries were required to demonstrate adequate levels of data privacy for data transfers to occur. Finally, a standing committee of national data privacy authorities, the Article 29 Working Party, was created to advise the European Commission on data privacy issues, to promote harmonized enforcement, and to evaluate the adequacy of privacy protection in non-EU countries'. This was still a Directive, a legislative act that articulated a goal to work towards to by member states based on the freedom to enact their own legislations. The GDPR, however, is a regulation with binding force across the EU. In other words, in the 23 years between the directive and the regulation, it became amply clear on the basis of multiple acts of omission and commission in the EU and outside, from the need for a uniform data sharing regime in the context of Schengen, to the Snowdon revelations and the reality of mass surveillance and the power of big data companies that 'data' had indeed become the commodity *par excellence*, generated by transactional consumers and central to profits of a handful of tech companies in the global economy.

---

[2] Valentini, C. (2017), European Commission: Directorate-General for Justice (Data Protection Division) (458–461), in Schintler, L.A. & McNeely, C. L. (Eds.), Encyclopedia of Big Data, Springer International Publishing, Switzerland.

[3] Newman, L. R. (2008), The EU Data Privacy Directive, Chapter 4 (74–98), in Newman, A. L. (Ed.), Protectors of Privacy: Regulating Personal Data in the Global Economy, Cornell University Press, Ithaca and London.

The Schrems Incident and Transborder Data Flows between the United States and EU

A key issue that contributed to the framework of the GDPR was the need to protect personal data in transborder data flows, which is key to trade in a global economy. The EU had recognized countries that maintained adequate levels of protection for personal data and that included the United States, that in 2019 accounted for US$1.1 trillion in trade and services with the EU including exports of $468 billion and imports of $598 billion (USTR 2020).[4] While the US's data protection regime was weak, the Safe Harbour scheme was established in the United States under the aegis of the US Department of Commerce and the US Federal Trading Commission and based on the voluntary self-certification of companies on data protection principles. This resulted in 3,200 US-based companies signing up. However, the Snowdon revelations of widespread surveillance and the culpability of Silicon Valley tech companies in the voluntary disclosure of data traffic to agencies such as the National Security Agency (NSA) linked to the Planning Tool for Resource Integration, Synchronization, and Management (PRISM) mass surveillance initiative, led to the EU redoubling their case for global security for personal data transfers from the EU. It was also the result of the Austrian data activist Max Schrems' legal challenges against Safe Harbour that was based on his own experience with big data companies, in particular Facebook (Kuchler 2018).[5] While studying as an exchange student at Santa Clara University, Silicon Valley, he attended a class taught by a Facebook employee that highlighted its ubiquitous data collection of its consumers. Schrems, using the framework of the EU's Data Privacy Directive, requested Facebook to send him all the data that they had on him and received a 1,200-page pdf document under 57 categories with extensive information on rejected friend requests, data that he had deleted, data on those 'defriended' and had 'poked' prompting him to take this up with the Irish Data Protection

---

[4] European Union, Office of the United States Trade Representative. Available at: https://ustr.gov/countries-regions/europe-middle-east/europe/european-union#:~:text=U.S.%20Goods%20trade%20(exports%20plus,was%20%24184%20billion%20in%202019
[5] Kuchler, H. (2018), Max Schrems: The man who took on Facebook and won, Financial Times, 5 April. Available at: https://www.ft.com/content/86d1ce50-3799-11e8-8eee-e06bde01c544

Commissioner, a country in which the Facebook subsidiary that administered all its European users was located (Pidd 2011). Maria Tzanou (2017:555)[6] has highlighted the EU's concerns with PRISM and privacy, with the rights of EU citizens to challenge surveillance measures in the United States and contesting how and what personal data can be used for. In 2016, a new framework for transatlantic data flows, the EU–US Privacy Shield was established although it is clear that US authorities maintain the right to be opaque on the matter of the interceptions of personal data transfers from the EU irrespective of the incorporation of new rights of redress and investments in an Ombudsperson, thus maintaining their right to the continuation of American Exceptionalism. As Tzanou (563) has observed while the lines of information flows between the EU and the US have been drawn for specific investigations related to data violations 'the Ombudsperson will neither confirm nor deny whether the individual has been the target of surveillance not will the Ombudsperson confirm the specific remedy that was applied'. It is possible that the Consumer Online Privacy Rights Act (2019), with its accent on the 'rights' of every US citizen over their personal data, and that establishes strong standards on the collection, use, sharing, and protection of personal data by companies such as Facebook will strengthen the protection of personal data in the United States thus making it easier for trans-border data flows between the EU and the United States although whether this Act will remain subservient to surveillance regimes in the United States remains an outstanding issue. Schrems' 23rd complaint to date on the limitations of the Privacy Shield was upheld in the European Court of Justice.

EU Sanctions and Enforcement

The EU's commitment to competition and data privacy has been relentlessly exemplified by the vast number of legal actions taken by the Commission's Competition Commissioner Margrethe Vestager against the tech giants for infringing the GDPR and other related EU related Acts related to privacy, data transfers, and competition. Google, for

---

[6] Tzanou, M. (2017), European Union regulation of transatlantic data transfers and online surveillance (545–565), Human Rights Law Review, 17.

example, has been fined annually between 2017 and 2019 for its anti-competitive practices, the abuse of its position in the 'search' space, the use of its Android system to strengthen its market dominance, and for placing 'restrictive clauses in their contracts with third-party websites' totalling €8.25 billion (EC: Antitrust Commission).[7] Facebook was fined €110 million for providing incorrect information on its merger with WhatsApp, while Amazon could be fined up to 10% (€19 billion) of its annual turnover for abusing its dominant power in online retail in an ongoing investigation (see BBC: Amazon charged).[8] The Digital Services Act and the Digital Markets Act will further the regulatory embrace over what has been described as 'attention utilities' (Harris 2020)[9] and will force the large tech giants to be responsible for online content, become visibly more transparent and will restrict their current forms of anti-competitive behaviour in their role as 'gatekeepers'. These Acts will apply to intermediary services, hosting services, and large and very large (platforms that reach 45 million consumers in Europe) (EU: The Digital Service Act).[10] The key objectives of these two Acts will be to create a safer digital place for online users and to create a level playing field both in the EU and at a global level. It has been mooted that recalcitrant firms can, as per these laws, be broken up and fined up to 10% of their annual turnover. Harris (2020)[11] has put forward the view that 'the EU should create a new corporate classification for large, dominant social platform businesses that have created vital public digital infrastructure. These "attention utilities" should be required to operate in the public interest, according to rules and licences that guide their business models ... Attention utilities should be required to obey limits on data extraction and message amplification practices that drive polarisation, and should be required to protect children'.

---

[7] Antitrust: Commission fines Google E1.49 billion for abusive practices in online advertising, European Commission. Available at: https://ec.europa.eu/commission/presscorner/detail/en/IP_19_1770

[8] Amazon charged with abusing EU competition rules (2020), 10 BBC, November. Available at: https://www.bbc.com/news/business-54887650

[9] Harris, T. (2020), EU should regulate Facebook and Google as 'attention utilities', Financial Times, 1 March. Available at: https://www.ft.com/content/abd80d98-595e-11ea-abe5-8e03987b7b20

[10] The Digital Services Act: Ensuring a safe and accountable online environment, EC. Available at: https://ec.europa.eu/info/strategy/priorities-2019-2024/europe-fit-digital-age/digital-services-act-ensuring-safe-and-accountable-online-environment_en

[11] Harris, Ibid.

So what are some of the key regulatory principles expressed in the GDPR that are relatively novel and that are aimed at strengthening the data privacy of EU citizens and the seeding of these principles in the global regulation of data platforms?

The Right to Be Forgotten

One of the most striking principles expressed in the GDPR is a person's 'right to be forgotten'. This right needs to be assessed within a context in which the voluntary, involuntary, and at times deceptive capture/appropriation of personal information has become the basis for a multibillion-dollar data processing empire controlled by a handful of global tech companies. The fact that information of a personal kind can remain in circulation online long after the demise of the person in question does highlight a complex of issues including the right to erasure of personal information as well as issues linked to identity fraud. However, the 'right to be forgotten' also applies to those who are physically present and have a life online but who would like to erase information in public circulation that may have been relevant in the past but not in the present. For example, a person may have had a run-in with the police that was reported in a newspaper, resolved offline, or through sentencing (limitation period) but that remains available online in perpetuity in different jurisdictions around the world. The issue here is with data processors and controllers, search engines such as Google and social media sites such as Facebook who have been loathe to 'erase' such data precisely because their terms of use with users do not include contractual obligations on their part on any limits and duration of their use of any given person's personal data/metadata.

The right to be forgotten is a relatively new human right and as such there is little historical jurisprudence or a body of thought on what is a complex human right. This right has been contested, not just by Big Tech companies who see it as yet another attempt to dampen their business plans but also by freedom of expression advocates such as the UK-based NGO Article 19 who have argued that the right to be forgotten cannot be an absolute right but a right that is dealt with in context where the circumstantial evidence points to the clear benefits from erasure. Their argument based on the need to balance the right to be forgotten with the freedom of expression and the right to information is certainly applicable in cases where the erasure of publically available records is criminally motivated and the means to hide information of say someone who wilfully bankrupted clients on the basis of, for example, a fraudulent investment scam such as a 'ponzi' scheme. In other words, information that needs to be made available precisely because of its abiding public interest.

Article 19 (2016:2)[12] points out that 'A strict seven-part test for balancing the right to freedom of expression and the "right to be forgotten" should be applied, taking into consideration:

- Whether the information in question is of a private nature;
- Whether the applicant had a reasonable expectation of privacy, including the consideration of issues such as prior conduct, consent to publication or prior existence of the information in the public domain;
- Whether the information at issue is in the public interest;
- Whether the information at issue pertains to a public figure;
- Whether the information is part of the public record;
- Whether the applicant has demonstrated substantial harm;
- How recent the information is and whether it retains public interest value'.

While Article 19's interpretation requires consideration, it would seem that they do not adequately address the question—whose freedom of expression? since we are not dealing with legacy media but ubiquitous media characterized by the exercise of major control over personal data—the right to control, limit, and censor speech by a handful of tech companies. While it is important to remember that the right to freedom of speech (First Amendment Rights) for private companies has been vigorously supported by corporate lobbyists in the United States, it is also important to interrogate the right to be forgotten within a context characterized by information asymmetries, data capture by infinite sensors,

[12] Article 19 (2016), 'The Right to be Forgotten': Remembering the Freedom of Expression (1–36), London. Available at: https://www.article19.org/resources/policy-brief-the-right-to-be-forgotten/

limitless storage, and processing capacities and global communications. As Ausloos (2020:18)[13] has observed, a group selfie uploaded on a social media site can become the basis for infinite mining at a variety of levels:

> The picture itself contains personal data of several individuals: faces, potentially 'sensitive data' (Arts 9–10 GDPR), metadata including timestamp and location, type of camera. Once uploaded, all of this information is accessible by the social network operator, by the uploader's 'friends', by connected third-party apps, advertisers, etc. Each of these entities pursues different interests and might further process the personal data in many different ways. Other individuals, for example, might reshare the picture; the social network operator might perform facial recognition on the pictures, identifying and linking the people in it and use it for fine—tuning its search algorithms and profile templates. Connected apps might use the picture (and/ or its metadata) to infer data that could then be repackaged and transferred to yet other parties. And so on.

The first ruling on the right to be forgotten followed a complaint by a Spanish national Mario Costeja Gonzalez in 2010 made to the Spanish Data Protection Agency (AEPD) against a newspaper company *La Vanguardia* that had published details of the forced auction of his property as a result of social security debts owed by him to the State and the subsequent circulation of this information online on Google, Spain. While the European Court of Justice (ECJ) dismissed the case against the newspaper company since they had carried this item in the public interest, the role of Google Spain and Google Inc. as processor-controllers of personal information long after, as in this particular case, debts had been repaid, was interpreted by the courts as the preponderant power of search engines to create a list of results and hyperlinks that allowed data to be present long after the cessation of its relevance and/or interest to the public and to the complainant (CJEU Press Release 2014).[14] Based

---

[13] Ausloos, J. (2020), The Right to Erasure in EU Data Protection Law, Oxford Scholarships Online. Oxford University Press.

[14] Court of Justice of the European Union, Press Release No. 70/14, Luxembourg, 13 May 2014. Judgment in Case C-131/12, Google Spain SL, Google Inc. v Agencia Española de Protección de Datos, Mario Costeja González. Available at: https://curia.europa.eu/jcms/upload/docs/appl ication/pdf/2014-05/cp140070en.pdf

on the interpretation of two items of law, Article 14 of Directive 95/46/EC of the European Parliament and of the Council of 24 October 1995 on the protection of individuals with regard to the processing of personal data and on the free movement of such data and of Article 8 of the Charter of Fundamental Rights of the European Union, the ruling in para 88 made the following observation that 'the operator of a search engine is obliged to remove from the list of results displayed following a search made on the basis of a person's name links to web pages, published by third parties and containing information relating to that person, also in a case where that name or information is not erased beforehand or simultaneously from those web pages, and even, as the case may be, when its publication in itself on those pages is lawful' (Judgement of the Court 2014).[15] This ruling, in addition to making the right to be forgotten a legal right, affirms the fact that data processors/controllers such as Google are subject to EU Data Protection Law. Google received 845,501 right to be forgotten requests during 2014–2019 leading to the removal of 45% of 3.3 million links. The ECJ in a ruling in 2019 confirmed that this right is only applicable within the boundaries of the EU, a victory or reprieve as this ruling may be for Google and to advocates of freedom of expression who have worked hard to limit the scope of this right (Marsh 2019).[16] The fact that one can access sensitive/personal data that is no longer available in the EU in other jurisdictions will remain controversial until and unless this right is included in privacy laws being considered under national jurisdictions the world over. What is clear is that this controversy will simply have to be resolved especially in jurisdictions such as the United States where enumerated rights such as the freedom of speech is encapsulated in the First Amendment of the US Constitution while privacy is not protected under the US Federal Constitution (Werro 2018).[17] Search engines such as Google have argued that the right to be forgotten and

---

[15] Judgement of the Court (Grand Chamber), 13 May 2014. Available at: http://curia.europa.eu/juris/document/document_print.jsf?doclang=EN&text=&pageIndex=0&part=1&mode=DOC&docid=152065&occ=first&dir=&cid=667631

[16] Marsh, S. (2019), 'Right to be forgotten' on Google only applies in EU court rules, 25 September. Available at: https://www.theguardian.com/technology/2019/sep/24/victory-for-google-in-landmark-right-to-be-forgotten-case

[17] Werro, F. (2018), The right to be forgotten: The General Report-Congress of the International Society of Comparative Law, Fukuoka, July 2018 (1–35), in Werro, F. (Ed.), The Right to Be Forgotten: A Comparative Study of the Emergent Rights Evolution and Application in Europe, the Americas and Asia, Global Studies in International Law, Springer, Switzerland.

Article 8 of the Directive imposes impossible obligations given that data that is deemed 'sensitive' such as the sexual orientation of data subjects may be available on blogs, websites, profiles on social network sites that are continuously indexed making compliance a difficult prospect (see Padova 2019).[18] It is clear that in an era of data capitalism, private firms are involved in governance via their involvement in setting standards and norms on matters such as the right to be forgotten. The hybridization of governance is a direct consequence of governments recognizing the power of Big Tech curate and control vast amounts of personal and non-personal data and data-sets (Chenou & Radu 2019).[19] While the right to be forgotten vs. freedom of expression remains a key issue, there are others such as the right to be forgotten vs. the right to memory that have been articulated by the memory studies scholar Noam Tirosh (2017).[20] Arguing that the right to be forgotten relates to one's right to construct one's own narrative, it makes a case for this right to be explored within memory studies and in the context of our digital times characterized by over-investments in technologies that aid remembering rather than forgetting. Remembering is enhanced by '(a) Interactivity, as contemporary media enable end-users to actively design and transform their media environment alone or collaboratively with remote end-users; (b) Mobility, as they transcend space limits through the use of nonstationary devices; (c) Multi-mediality, as users have the potential to utilize written words, sounds, static and moving images, all or any at any given time; and (d) Abundance, as they carry the potential to access and deliver infinite amounts of information over endless communication channels' (648). There is also an argument that deletion and erasure will contribute to the perpetuation of the conceit of human perfection, of sanitized selves that fly against the complexity of self (Garcia-Murillo & Macinnes 2018).[21] The communitarian philosopher Amitai Etzioni (2015:122)[22] refers to

---

[18] Padova, Y. (2019), Is the right to be forgotten a universal, regional or 'glocal' right? (15–29), International Data Privacy Law, 9 (1).

[19] Chenou, J-M. & Radu, R. (2019), The 'right to be forgotten': Negotiating public and private ordering in the European Union (74–102), Business & Society, 58 (1).

[20] Tirosh, N. (2017), Reconsidering the 'right to be forgotten'—memory rights and the right to memory in the new media era (644–660), Media, Culture & Society, 39 (5).

[21] Garcio-Murillo, M. & Macinnes, I. (2018), Cosie Fan Tutte: A better approach than the right to be forgotten (227–240), Telecommunications Policy, 42 (3).

[22] Etzioni, A. (2015), Privacy in a Cyber Age: Policy and Practice, Palgrave Macmillan, New York.

the right to be forgotten as a 'hedged right' and not a generic right since it can be recalibrated to meet the needs of the communitarian, common good. Perhaps the most ambitious defence of the right to be forgotten is Viktor Mayer-Schonberger's (2011) *Delete: The Virtue of Forgetting in the Digital Age* in which a case is made for a move away from remembering as the default position in an era characterized by information overload to forgetting aided by social norms, laws, and technical architecture. From forgetting being the default position over the ages, today it is eternal 'external' memory that has become the default position as opposed to finite, biological memory. This is described in chapter 5 (128–168) where responses from *individuals* (abstinence and cognitive adjustment, *laws* (privacy rights and information ecology) and *technology* (privacy DRM and full contextualization) are presented as the means by which the expiry dates for personal data in external memory is institutionalized thus achieving the goal of forgetting. As Mayer-Schonberger (198)[23] has argued, 'Expiration dates are relatively modest on a number of implementation dimensions, making them comparatively easy to adopt. And yet they may be enough to stop and reverse the shift towards remembering and restore our capacity to forget, which is so central to what it means to be human'.

The right to be forgotten is closely linked to privacy and is a substantive enlargement of the rights of data subjects. The onus as expressed in Article 17 of the GDPR is on data controllers to act on requests to erase data without undue delay. Fundamentally though, this right is founded on a more comprehensive understanding of 'data' as the primary fuel for production and the creation of value in the digital age than was hitherto the case. Personal data has become the basis for a multi-billion industry, and, as such, the GDPR recognizes the centrality of data in informational capitalism as a new 'asset class' the bulk of which is produced through transactional user generation. Article 4(1) GDPR[24] defines personal data as: 'any information relating to an identified or identifiable natural person ('data subject'); an identifiable natural person is one who can be identified, directly or indirectly, in particular by reference to an identifier

---

[23] Mayer-Schonberger, V. (2011), Delete: The Virtue of Forgetting in the Digital Age, Princeton University Press, Princeton and Oxford.

[24] Art.4 GDPR Definition, Intersoft Consulting. Available at: https://gdpr-info.eu/art-4-gdpr/

such as a name, an identification number, location data, an online identifier or to one or more factors specific to the physical, physiological, genetic, mental, economic, cultural or social identity of that natural person.'

Personal data is both product and substance. It is this recognition of
power and knowledge imbalances in the digital age that is key to the
understanding of the rights of data subjects in the GDPR including
their right to be forgotten. As the critical information theorist Mark
Andrejevich (2015:11)[25] has argued 'A theoretical approach that does
not engage with these aspects of the continuing relevance of the ownership and control over infrastructure in the commercial logic of the online
economy seems ill-suited for conceptualizing the emerging role of the
capture and use of various kinds of data—including personal data—and
the power asymmetries that are thereby reinforced'. It is the economic di-
mension of personal data and its exploitation by Big Tech companies that
have led the EU to define the protection of personal data as a fundamental
right. Since cultures based on affective socialities are today both global
and local, all manner of data including intimate data have the potential to
become publically available. Pangrazio and Selwyn (2019:421–422)[26] describe personal data as 'data that users give to devices/systems, data that
devices/systems extract from users and data that devices/systems process
on behalf of users'. It is this power of Big Tech companies to control the
making, storage, and use of personal data that has led to the GDPR including in the responsibilities of data controllers. the need for them to receive prior consent from data subjects for the collection of data, for them
to inform data subjects on the use of their data, their right of access to
the data and their rights to rectify and cancel the data. Article 12 of the
GDPR makes a clear case for privacy clauses to be 'concise, transparent,
intelligible and easily accessible', thus forcing Big Tech companies to exercise transparency in their licensing contracts with users. The GDPR
recognizes the propensity for Big Tech companies to obfuscate—a word
that is described by Brunton and Nissenbaum (2015:1)[27] as '*the deliberate*

[25] Andrejevich, M. (2015), Personal data: Blind spot of the 'affective law of value'? (5–12), The Information Society, 31 (1).

[26] Pangrazio, L. & Selwyn, N. (2019), 'Personal data literacies': A critical literacies approach to enhancing understandings of personal data (419–437), New Media & Society, 21 (2).

[27] Brunton, F. & Nissenbaum. H. (2015), Obfuscation: A User's Guide to Privacy and Protest, MIT Press, Cambridge: Massachusetts and London: England.

addition of ambiguous, confusing, or misleading information to interfere with surveillance and data collection' (emp. authors). The emphasis on transparency and consent in the GDPR explicitly is anti-obfuscatory— and non-compliance entails substantial business costs.

The right to be forgotten in addition to the ARCO rights (right of access, rectification, cancellation, and opposition) is accompanied by another right—the right to portability of personal data. This right gives data subjects the right to move their personal data from one platform to another. Data processors are expected to provide this data in a standard, transposable, readable, machine learning format, transmit this to other platforms and delete all copies of this data found on their servers. Data processors are also expected to transfer this data without hindrance, meaning that they cannot obstruct or obfuscate such transfers, levy a fee, or use formats that are not easily translatable across data processors. Data portability enables choice, strengthens competition, format interoperability, helps the development of the Single Digital Market in the EU within a competitive environment, and gives data subjects the right to do as they please with what are personal cultural memories. Ursic (2018:59)[28] has interpreted data portability as an enhancement of the rights of data subjects. It enables four gateways that help strengthen the rights of data subjects:

'1. Establishing control over personal data transfers;
2. Establishing control over (re)uses of personal data;
3. Enabling better understanding of personal data flows and their complexity; and
4. Facilitating free development of personality and enhancing equality'.

Arguably, both the right to be forgotten and data portability rights are 'aspirational' rights meaning that while some case law is available, the jurisprudence on these rights is in its infancy. Both rights are difficult to operationalize in practice given the many locations that store transactional data that include both the public and private sectors. There remain substantive differences in legal interpretations of personal and non-personal

[28] Ursic, H. (2019), Unfolding the new-born right to data portability: Four gateways to data subject control (42–69), Scripted 15 (1).

data, what in the GDPR is referred to as 'provided data'. While the rights to data portability applies to actively and passively generated data by data subjects, 'inferred' data that is based on algorithmic analysis and assessed/manipulated via analytical techniques cannot be ported. A report from the European Network and Information Security Agency (ENISA) on data portability highlights the complexity of the gathering of personal data from 'open' as opposed to 'closed' networks. 'The fundamental technical challenge in enforcing the right to be forgotten lies in (i) allowing a person to identify and locate personal data items stored about them; (ii) tracking all copies of an item and all copies of information derived from the data item; (iii) determining whether a person has the right to request removal of a data item; and, (iv) effecting the erasure or removal of all exact or derived copies of the item in the case where an authorized person exercises the right' (Druschel, Backes, & Tirtea 2011:8).[29] Freedom of expression and free speech advocates continue to contest the right to be forgotten. Furthermore, data processors and controllers such as Google have been involved in contesting this right in courts highlighted by the dispute between the French privacy regulator, the National Commission for Computing and Liberties (CNIL) and Google and their attempt to get Google to erase all data related to a particular individual. The European Court of Justice ruled that Google could not be expected to erase any reference to this person outside of the EU.[30]

The EU has expressed the clearest intent to regulate Big Tech and it is the case that evolving laws in the EU are becoming the touchstone for regulatory laws in other jurisdictions around the world. In some ways, their lead is what is required for the rest of the world as Big Tech will simply have to, given the market power of the EU, negotiate with them on terms set by the EU. This could lead to the establishment of general principles that have universal applicability.

---

[29] Druschel, P., Backes, M., & Tirtea, R. (2011), The right to be forgotten: Between expectations and practice (1–22), European Network and Information Security Agency, Heraklion, Greece. Available at: file:///Users/uqpthom4/Downloads/The%20right%20to%20be%20forgotten%20-%20%20between%20expectations%20and%20practice.pdf

[30] Pidd, H. (2011), Facebook could face €100,000 fine for holding data that users have deleted, The Guardian, 21 October. Available at: https://www.theguardian.com/technology/2011/oct/20/facebook-fine-holding-data-deleted

4

Platform Regulation in the United States

Since the United States is home to most of the world's leading Big Tech companies and platforms that have exerted a dominant economic presence for over four decades, one could be forgiven for thinking that this growth in the depth and reach of these companies has been accompanied by an equivalent development in its governance and regulation. After all, there are regulatory regimes for all economically productive sectors and oversight mechanisms such as the Food and Drug Administration (FDA, food), the Federal Communications Commission (FCC, broadcasting), and the Federal Aviation Commission (FAC, aviation), among other such bodies. So why is it the case that large tech companies are beyond the purview of such regulation? Would it be the case that (1) this state of play and policy vacuum is the consequence of a deliberate policy adopted to ensure global competitive advantage, (2) that owing to the rapid pace of innovation in the digital it has been impossible for policymakers to establish rules of the game, (3) that the experience of digital saturation, ubiquity, and control of 'affective' labour and its 'game-changing' consequences at a variety of levels is only now being recognized across the political divide in the United States and in the context of hate speech, censorship, political and economic manipulations of consumer behaviour, (4) that the reason for this continuing vacuum is the tremendous money and lobbying power of these tech giants and, (5) this is as is because the figuring out of how to regulate the digital and data is profoundly contentious. The fact that Sir Tim Berners-Lee (2021),[1] an otherwise progressive on matters digital, has come out strongly in support of platforms (the principle that hyperlinks should not be monetized in the interest of net neutrality) in the ongoing battle in Australia aimed at ensuring that

---

[1] Berners-Lee, T. (2021), Treasury Laws Amendment (News Media and Digital Platforms Mandatory Bargaining Code) Bill 2020 Submission 46. Available at: file:///Users/uqpthom4/Downloads/Submission%2046%20-%20Sir%20Thomas%20Berners-Lee%20(2).pdf

Platform Regulation. Pradip Ninan Thomas, Oxford University Press. © Pradip Ninan Thomas 2023.
DOI: 10.1093/oso/9780192887962.003.0004

platforms compensate legacy media for carrying their content, illustrates the complexities regarding digital regulation. While maximum access to content, to sharing online, and restricting the monetization of 'fast' lanes as opposed to slow are general principles that have merit, network neutrality has already been breached multiple times through legal means and business practices. In this context, there simply is the need for a clear understanding of net neutrality from the perspective of the public good—for example, the need for unrestricted access to government websites that host evidence-based information on COVID-19 pandemic, as opposed to websites and unregulated platforms that host contrary meanings.

While each of the five reasons explained previously may contribute to our understanding the state of play related to digital regulation in the United States, when compared with the European Union (EU), it is clear that the 'will' to regulate is of recent vintage and is linked to a large extent to domestic anxieties related to the 2016 elections—a watershed moment in the recognition of the extent of the manipulation of affective behaviours online along with the growing political unease of the powerful role played by Big Tech intermediaries in shaping the terms for economic competition in the United States. There is an ambivalence linked to the regulation of Big Tech because both the Republicans and Democrats have benefited from and have been challenged by the operations of Big Tech companies. Big Tech have donated liberally to both sides of politics in the United States although the major beneficiary of donations in 2020 was the Democrat, Joe Biden. Alphabet's political donations in 2020 was $21 million, Microsoft's $17 million, Amazon's $8.9 million, Facebook's $6 million, and Apple $5.7 million (Cao & Zakarin 2020).[2] Worsdorfer (2020:195)[3] has highlighted employee flows between the government and Big Tech companies in the United States, an illustration of nexus politics and conflicts of interest. 'This power is not only used to purchase start-ups and other (rival) companies; it has also been used for lobbying and rent-seeking purposes: Data published by the Google Transparency Project and the Campaign for Accountability show that an increasing

---

[2] Cao, S. & Zakarin, J. (2020), Big tech and CEOs poured millions into the elections. Here's who they supported, Observer, 11 February. Available at: https://observer.com/2020/11/big-tech-2020-presidential-election-donation-breakdown-ranking/
[3] Worsdorfer, M. (2020), Ordolibralism 2.0: Towards a new regulatory policy for the digital (191–215), Philosophy of Management, 19.

number of employees has moved back and forth between Alphabet and the various US government agencies—pointing towards the so-called "revolving door phenomenon", known from the finance industry.

Following the riots in Capitol Hill in January 2021, Big Tech halted political donations while Airbnb, AT&T, Cisco, Comcast, Intel, and Verizon have stopped giving to those who challenged the election results (Gold & Fried 2021).[4] Following the 2020 elections, Trump accused Facebook and Twitter of censoring his side of politics although he used Twitter very effectively to distribute 'fake news' and misinformation on an election that he claimed was 'stolen'. At the same time, the US government has very assiduously supported the operations of Big Tech on the global stage and resisted any attempts to regulate Big Tech via taxation policies such as in France, the EU, and India and compensation policies such as in Australia. They have also supported the data imperialism that Big Tech have been involved in and resisted moves to bring Big Tech under the embrace of data privacy and data localization in jurisdictions such as the EU and India. Three policy moves in India that support data localization— the Personal Data Protection Bill, E-Commerce Policy, and Guidelines for Intermediaries have been contested by the US Trade Representative (USTR) for their perceived impact on US tech companies and a threat to initiate a Section 301 review followed by sanctions/imposition of tariffs for Indian exports is under consideration (see Raghuraman 2019; Basu, Hickok, & Chawla 2019).[5]

Anti-trust

At the very centre of this ambivalence is the continuing questioning of the validity and jurisprudential worth of legal actions such as 'anti-trust'

[4] Gold, A. & Fried, I. (2021), Tech companies press pause on political donations, Axios, 12 January. Available at: https://www.axios.com/capitol-siege-tech-political-donations-8823b80d-e205-40a1-9812-d24f9dc03d25.html

[5] Raghuraman, A. (2019), Don't let tech policy disrupt the US-India trade, Atlantic Council, 10 December. Available at: https://www.atlanticcouncil.org/blogs/new-atlanticist/dont-let-tech-policy-disrupt-the-us-india-trade-deal/

Basu, A. Hickok, E., & Chawla, A. S. (2019), The localisation gambit: Unpacking policy measures for the control of data in India (1–90). Centre for Internet and Society, 19 March. Available at: https://cis-india.org/internet-governance/resources/the-localisation-gambit.pdf

that are seen to be legacies of the era of brick and mortar not compatible with the era of data. Nicholas Petit (2020:187),[6] for example, in his book *Big Tech and the Digital Economy* suggests that the preservation of 'rivalry' that is at the core of anti-trust should be abandoned in the context of the non-rival economics of digital markets. According to him, destabilizing already 'tipped' markets dominated by a monopoly or oligopoly providers providing extensive services to consumers should be avoided and instead competition in adjacent, untipped markets characterized by uncertainty and instability encouraged. 'The main function of antitrust in digital markets should not be to promote rivalry—it cannot be presumed to be socially efficient—but a *pressure* equivalent to it. In particular, by maintaining pressure on monopoly rents, antitrust creates an uncertainty equivalent to rivalry that produces powerful incentives on established big tech firms to invent new products and introduce market-shifting innovations. Antitrust remedies should be introduced to promote uncertainty in big tech firms' monopoly markets when it has disappeared as a result of bad conduct and structural market failures or features like network effects, increasing returns to adoption, and market tipping'. Anti-trust as a moral position and action supportive of consumer interests has been critiqued by anti-trust scholars such as Herbert Hovenkamp (2005)[7] who see it purely in terms of its utilitarianism. The Chicago School of anti-trust scholars are of the view that the key objective of anti-trust was to ensure greater efficiencies in the market and the protection of competition along with keeping judicial interventions at a minimum and trust in the market as the final arbiter. Theirs was an argument that gave weight to 'allocative and productive efficiencies' and was based on empirical estimates of 'the relative costs of overinclusion and underinclusion of antitrust rules' (Page 1989:1226).[8] Arguably, their approach to anti-trust favoured monopoly behaviours and influenced the Federal Trade Commission's (FTC) and the views of other bodies involved in anti-trust actions in the United States. Sokol and Comerford

---

[6] Petit, N. (2020), Big Tech and the Digital Economy: The Moligopoly Scenario, Oxford Scholarships Online, Oxford University Press.

[7] Hovenkamp, H. (2005), The Antitrust Enterprise: Principle and Execution, Harvard University Press, Cambridge, MA.

[8] Page, W. H. (1989), The Chicago School and the evolution of antitrust: Characterization, antitrust injury and evidentiary sufficiency (1221–1308), Virginia Law Review, 75 (7).

(2017:296)[9] have argued against using of anti-trust to rein in Big Data companies by taking the position that the monetization of consumer behaviour should not to be suspected but valued as 'economically-rational, profit-maximising behaviour' that has 'obvious consumer benefits' that includes free products and subsidised services. These examples of anti, anti-trust writings, highlight rationales for protecting monopolies that range from arguments in favour of consumer benefits, natural economic behaviours such as predatory pricing to those advancing the need to restrict inefficient competition from natural monopolies. There are also those who err on the side of Big Tech self-regulating—an example of this is Facebook's proposed 20-member strong Review Board that has been mandated to deal with privacy, hate speech, and harassment although as the social media critic Siva Vaidhyanathan (2020)[10] has observed, this body:

> will hear only individual appeals about specific content that the company has removed from the service—and only a fraction of those appeals. The board can't say anything about the toxic content that Facebook allows and promotes on the site. It will have no authority over advertising or the massive surveillance that makes Facebook ads so valuable. It won't curb disinformation campaigns or dangerous conspiracies. It has no influence on the sorts of harassment that regularly occur on Facebook or (Facebook-owned) WhatsApp. It won't dictate policy for Facebook Groups, where much of the most dangerous content thrives. And most importantly, the board will have no say over how the algorithms work and thus what gets amplified or muffled by the real power of Facebook.

One of the more interesting solutions linked to self-regulation has been advocated by Jack Balkin (2016)[11] through the concept of 'information fiduciaries'. He has argued in favour of a solution that reconciles the First

---

[9] Sokol, D. D. & Comerford, R. (2017), Does antitrust have a role to play in regulating Big Data (296–316), in Blaird, R. D. & Sokol, D. D. (Eds.), The Cambridge Handbook of Antitrust, Intellectual Property and High Tech, Cambridge/New York/Melbourne/New Delhi/Singapore.

[10] Vaidhyanathan, S. (2020), Facebook and the folly of self-regulation, Wired, 5 September. Available at: https://www.wired.com/story/facebook-and-the-folly-of-self-regulation/

[11] Balkin, J. M. (2016), Information fiduciaries and the first amendment (1183–1234), UC Davis Law Review, 49 (4). Available at: file:///Users/uqpthom4/Downloads/SSRN-id2675270.pdf

Amendment guarantees of Free Speech with regulation (1183) and in which new online business involved in the collection, use, storage, and sale of personal data should be recognized as information fiduciaries like doctors and lawyers who are expected to keep information of their clients confidential and who exercise a duty of care and duty of loyalty to their clients. If indeed Big Tech companies such as Facebook are seen as public utilities rather than a private company, their relationship with their customer base could be viewed from within their role as an information fiduciary.

It is clear that anti-trust means different things in different jurisdictions such as the EU where consumer welfare and competition have been key to anti-trust actions against Big Tech companies. However, even in the United States where the Department of Justice (DoJ), the Supreme Court, and the FTC have presided over many hundreds of mergers and acquisitions by the Big Tech companies between 2009 and 2019, there is evidence of innumerable actions at a Congressional and other levels to curb the power of Big Tech. These include anti-trust lawsuits filed against Google for its anti-competitive behaviour by 30 states in the United States in December 2020, by the FTC and 48 states against Facebook also in December 2020 and that was preceded by similar lawsuits filed by the Justice Department and 11 states in October 2020 against Google (Kari 2020).[12] These suits against the 'de facto exclusivity' of these Big Tech firms have occurred after a gap of 22 years—1998, when Microsoft was accused of anti-competitive behaviour under the Sherman Act (1890). In August 2020, a Democrat subcommittee released a 400-page report on the monopoly power of Big Tech while a Republican subcommittee also published their report that same month (Edelman 2020).[13] And in the context of the storming of Capital Hill in Washington in January 2021, there have been moves to amend Section 230 of the Communication Decency Act (1989) that gave blanket immunity to tech companies for user-generated content on their site (Thornhill 2021).[14] The recognition

---

[12] Kari, P. (2020), 'This is big': US lawmakers take aim at one untouchable big tech, The Guardian, 19 December. Available at: https://www.theguardian.com/technology/2020/dec/18/google-facebook-antitrust-lawsuits-big-tech

[13] Edelman, G. (2020), Congress unveils its plans to curb Big Tech's power, Wired, 10 August. Available at: https://www.wired.com/story/congress-unveils-plan-curb-big-tech-power/

[14] Thornhill, J. (2021), The tech platforms are not entirely to blame for Washington unrest, The Financial Times, 7 January. Available at: https://www.ft.com/content/ef64b160-5f01-404a-bb39-d013fde808ca

of the power of Big Tech and its approximation of public utility functions has led to thinktanks such as the Brookings Institute making a case
for a new Digital Platform Agency on a par with the FCC and FTC and
involved in digital policy oversight inclusive of risk management, res-
toration of common law principles such as the duty of care, and agile
regulation (Wheeler et al. 2020).[15]

Privacy

Apart from anti-trust actions against the anti-competitive actions of Big
Tech companies, there are also moves to create a federal privacy law that is
to some extent like the EU's General Data Protection Regulation (GDPR).
In lieu of a comprehensive, up-to-date federal privacy law, a number of
states in the United States have either mulled over or established their
own privacy laws including the California Consumer Protection Act
(CCPA), Illinois' biometric data laws, along with pending privacy legislations in the states of New York, Massachusetts, Maryland, Hawaii, and
North Dakota that covers both consumer rights and business obligations
(see Rippy 2021 for an updated list).[16] The Republicans have tried hard
to deny the establishment of 'stricter' privacy laws at the state level in fa-
vour of 'federal pre-emption', a federal law that would overrule other laws.
There have also been with various levels of success, attempts to intro-
duce privacy-linked Acts and Bills to the Senate including the Consumer
Online Privacy Rights Act, United States Consumer Data Privacy
Act, Data Protection Act, Mind your Own Business Act, Filter Bubble
Transparency Act, Social Media Privacy Protection and Consumer
Rights Act, Do Not Track Act, Designing Accounting Safeguards to
Help Broaden Oversight and Regulations on Data (DASHBOARD) Act,
Browser Act, and the Commercial Facial Recognition Privacy Act (Kelly

---

[15] Wheeler T., Verver, P., & Kimmelman, G. (2020), The need for regulation of big tech beyond
antitrust, Brookings, 23 September. Available at: https://www.brookings.edu/blog/techtank/
2020/09/23/the-need-for-regulation-of-big-tech-beyond-antitrust/

[16] Rippy, S. (2021), US state comprehensive privacy law comparison, International Association
of Privacy Professionals (IAPP), 4 January. Available at: https://iapp.org/resources/article/state-
comparison-table/

2020; Fazlioglu 2019).[17] The Consumer Online Privacy Bill of Rights (2012) is viewed as a foundational base for all-encompassing privacy law fit for the digital age. There have been federal privacy-related acts in the past including the US Privacy Act (1974) that focussed on outlining restrictions related to private data held by government agencies, data confidentiality in the Health Insurance Portability and Accountability Act (1996), the Children's Online Privacy Protection Act (COPPA), 2000, and the protection of non-public personal information in the Gramm-Leach-Bliley Act related to banking and financial law (Green 2020).[18] Over the last decade, close to 50 states have passed laws related to the notification of data breaches. In spite of all these legislations, the US consumer and primary generator of data is not covered by a law that accounts for user-generated data through web searches, social media, e-commerce, and smartphone apps.

What the above description of privacy initiatives in the United States highlights is definitely not a lack of it but the fact that they are varied, inconsistent, and incomplete resulting in a loss of clarity for consumers on their rights to data, the nature of consent and its use by companies on the one hand and the obligations and responsibilities of companies that have monetized personal data. Bennet and Raab (457:2020)[19] have argued that there have been 'four important and, to an extent, interrelated developments in regulatory conceptualization that serve to modify or extend the privacy paradigm, or at least test its resilience. One is the rise of *accountability* as a regulatory and self-regulatory philosophy and technique. A second is the (re-)discovery of *ethics* and its potential shaping of principles, rules, and behavior in the processing of personal data. A third is the increased prevalence of *risk-based and harm-based understandings* of privacy regulation, and their associated methodology, alongside the more *rights-based* philosophy. A fourth is the conceptualization of

---

[17] Kelly, M. (2020), All the ways Congress is taking on the tech industry, The Verge, 2 March. Available at: https://www.theverge.com/2020/3/3/21153117/congress-tech-regulation-privacy-bill-coppa-ads-laws-legislators
Fazlioglu, M. (2019), Tracking the politics of US privacy legislation, International Association of Privacy Professionals (IAPP), 13 December. Available at: https://iapp.org/news/a/tracking-the-politics-of-federal-us-privacy-legislation/
[18] Green, A. (2020), Complete guide to privacy laws in the USA: Compliance and regulation, Varonis, 29 March. Available at: https://www.varonis.com/blog/us-privacy-laws/
[19] Bennett, C. J. & Raab, C. D. (2020), Revisiting the governance of privacy: Contemporary policy instruments in global perspective (447–464), Regulation & Governance, 14.

privacy as having a *social value*, as well as being an individual right'. It is clear that when considering the experience of privacy in the United States that these four principles, although cited in jurisprudence, nevertheless are not part of a consistent framework for dealing with the many issues raised by privacy in the country that is home to most of the world's top tech giants.

Mark Zuckerberg, the CEO of Facebook, has argued that privacy has become an outmoded concept given that 'sharing' of data has become a contemporary social norm (Sherman 2020).[20] This dismissal of privacy also hides the extent to which machine intelligence and algorithmic orderings have become fundamental to the operations of search engines and social media firms. Indeed, algorithms play a fundamental role in computational generations of knowledge and yet in the midst of all the debates on privacy, the regulation of algorithms rarely if ever is considered (Yeung 2018).[21] The role played by algorithms in promoting 'design-based governance' (Gritsenko & Wood 2020)[22] is the way in which control is exercised in data-driven societies such as in the United States by tech companies although the need for oversight over the exercise of algorithmic power has not featured consistently in data governance frameworks in the United States. Part of the reason for this state of affairs is the partisan politics practiced in the United States and the ways in which the different ends of the political spectrum have defined privacy. So the Online Privacy Act that would have led to the establishment of a new Digital Privacy Agency and provide user rights including the rights of the user to delete and correct data, data portability, and private rights to actions has been delayed because of opposition to private rights to actions. Republicans and Democrats have been at loggerheads with respect to a pre-emptive Federal Privacy law that would supersede state-based laws highlighting the fact that bipartisan laws are difficult to establish in a highly polarized political climate. Arguably, this state of affairs has

---

[20] Sherman, J. (2020), Oh sure, Big Tech wants regulation—on its own terms, Wired, 28 January. Available at: https://www.wired.com/story/opinion-oh-sure-big-tech-wants-regulation-its-own-terms/

[21] Yeung, K. (2018), Algorithmic regulation—a critical interrogation (505–523), Regulation & Governance, 12

[22] Gritsenko, D. & Wood, M. (2020), Algorithmic governance: A modes of governance approach (1–18), Regulation & Governance, Available at: https://doi-org.ezproxy.library.uq.edu.au/10.1111/rego.12367

been to the advantage of Big Tech companies, some of whom who have over the space of two decades used these divides to their advantage and have become either $trillion or close to $trillion companies. This can be illustrated with the example of the Fair Information Practices Principles outlined by the FTC in the late 1990s and that mandated commercial websites to include information on Choice (individual consent), Notice (disclosure of their uses of personal information), data security, and so on. However, the operationalization of these principles has largely been left to these companies resulting in a lack of clarity at the user level on a range of issues related to privacy rights including informed consent. This example reinforces the view that in the United States, the self-regulation of privacy has not been to the benefit of the ordinary data subject. Esteve (2017:44)[23] highlights one example from many that stems from this lack of oversight and its implications for users in the United States. 'The "Choice" Fair Information Practice Principle requires companies to offer consumer choices as to how their personal identifying information is used beyond the use for which the information was provided. The Federal Trade Commission has often required companies to make modifications to their privacy policies to better notify users that their personal information is being collected, used, and shared, as these practices are considered to be unfair acts that violate the Federal Trade Commission Act. In Re Google Inc, the Federal Trade Commission charged Google with violating the Federal Trade Commission Act in connection with Google's launch of its social networking tool 'Google Buzz' because Google led users to believe that they could choose whether or not they joined the Buzz network but the options for declining were essentially ineffective'. Big Tech, however, are not united in their opposition to privacy and this adds another layer to the politics of privacy in the United States. While search engine and social media sites such as Google and Facebook have benefited from the lack of privacy, tech companies such as Apple have launched their own privacy campaigns such as Apple's 'A Day in the Life of Your Data' that tracks user data across websites and Apple's App tracking Transparency (ATT) policy that tightens use of information

---

[23] Esteve, A. (2017), The business of personal data: Google, Facebook and privacy issues in the EU and the USA (36–47), International Data Privacy Law, 7 (1).

from websites and apps for advertising purposes and that has been contested by Google and Facebook (Combette 2021).[24]

Platform Regulation: The FCC vs. the FTC

From the above rather cursory reading of data regulation in the United States, it is clear that while this country has been the bedrock for innovative and disruptive technologies and trends related to the digital that have had global consequences, in particular, the ubiquity of a global mode of production based on the generation, curation, storage, retrieval, manipulation, and sale of personal and non-personal data, the primacy of technological shapings have not been accompanied/augmented by concerted attempts to establish normative frameworks for data management. There have been multiple, varied efforts to regulate this economy although they seem scattered, silo approaches when what is required is a comprehensive approach to dealing with data as the currency for all productive activity by both data subjects and data controllers. Big Tech's ownership and control of the data landscape—its material infrastructures and the immaterial flows of data across all productive spheres of the economy, leisure, sociality, and politics is unprecedented and is a level up from previous generations of legacy media corporations in the United States that controlled large swathes of media markets across a variety of media sectors such as News Corp., Disney, and Viacom. Although these traditional corporates remain strong, their modus operandi and ability to control over the flow of ideas and meaning have been disrupted by Big Tech intermediaries that have radically reinvented the terms for the generation and valuation of data along with control over the generative power of data and the conduits, hardware, and software through which this data flows. Their parasitic appropriations of content/data and business models based primarily on exchange and only secondarily on production has resulted in an upending of traditional approaches to content protection such as intellectual property in the wake of the sharing economy and disembedding of

---

[24] Combette, C. (2021), Preparing our partners for Apple's iOS14 policy updates, Google Ads, 27 January. Available at: https://blog.google/products/ads-commerce/preparing-developers-and-advertisers-for-policy-updates/

the corporate 'author' and re-embedding within a political economy that is both uncertain and unpredictable. The ability of legacy media to attract advertising revenues has been radically altered by Big Tech companies in particular Google and Facebook. The political economy of search and sociality has major implications not just for the leisure industries but for user-consumers whose data interactivities across multiple devices are shaped by frameworks provided by Big Tech.

One of the salient issues with data regulation in the United States is whether the current regulatory dispensation and involvement of bodies such as the Federal Communications Commission, Federal Trade Communications, and other bodies have been able to do justice to the regulatory oversight required for a fast-changing, data-driven, environment. For example, in lieu of a strong regulatory environment, platforms such as Facebook periodically attempt 'self-regulation' although it is often the case that such steps result from the evidence of the misuse of this platform for political purposes or the manipulation of its users for economic gain. Barret and Kreis (2019:3)[25] refer to what they term 'platform transience' meaning the organizational means employed by platforms to refine/change policies, procedures, and affordances in response to normative pressures or the perceived drawbacks or the need to better services. So Facebook has introduced changes to its policy related to political advertising by removing/adding features often without any input from those who have been negatively affected. Twitter's removal of Trump in the light of the post-electoral chaos could be seen as a knee-jerk reaction by a company that benefited from Twitter being used as a megaphone. From a regulatory perspective, platform transience as a self-regulatory process is hugely problematic since such adjustments to processes and procedures impact vast numbers of users. The accountability of in-house changes needs to be viewed against the scale and reach of these platforms. If such changes to policies and procedures say in banking warrant regulatory oversight in the offline world, surely there is a need for similar regulatory oversight for online platforms whose audience/consumer/user base is often accounted for in the billions (Google 4 billion and Facebook 2.8 billion). Google's taken for granted 'participatory surveillance' is a search

---

[25] Barrett, B. & Kreiss, D. (2019), Platform transience: Changes in Facebook's policies, procedures and affordances in global electoral politics (1–22), Internet Policy Review, 8 (4).

conditionality that is based on user enabling of cookies (Rogers 2018:9).[26] The FCC's inability to ensure competition in legacy media in the United States and for that matter support 'public interest' media is especially problematic since their regulatory remit under the Communications Act (1934) is vast and includes the allocation of spectrum, regulate telecommunications, approve new consumer devices and approve millions of dollars of subsidies. This Act, which was established in the age of legacy media, is now being interpreted ex-ante and in a prescriptive fashion to accommodate the complexities of data-driven innovation. As Ohlhausen in a critique of the FCC (2015:209)[27] has observed 'Starting with the 1913 Kingsbury Commitment, through the 1996 Telecommunications Act and its subsequent implementation, Congress and the FCC constructed a regulatory framework that distinguishes among services based on their physical platform, business model, and geographic characteristics—distinctions that are increasingly irrelevant. Consequently, when considering the converging technologies and overlapping business models of an IP-based world, the FCC has struggled to deploy its prescriptive ex ante regulation tool in a manner that is both effective and legally sustainable'. The FCC has also been shaped by political interests (the repeal of Obama-era net neutrality in 2017) and the power of private sector media actors, the latter who have ensured limited competition in the media landscape. Brent Skorup (2016)[28] in the online version of the journal *National Affairs* highlights one such case of collusion resulting in the stymieing of competition and consumer choice '... one wireless company, LightSquared, spent billions of dollars converting satellite spectrum for use as mobile broadband, in the hopes of competing with AT&T and Verizon. The FCC tentatively encouraged that costly process for a few years before rescinding crucial permissions under intense political pressure. The bureaucratic shift immediately bankrupted LightSquared and deprived Americans of the benefits of another major wireless operator'.

[26] Rogers, R. (2018), Aestheticizing Google critique: A 20-year retrospective (1–13), Big Data & Society, doi: 10.1177/2053951718768626

[27] Ohlhausen, M. K. (2015), The FCC's knowledge problems: How to protect consumers online (203–234), Federal Communications Law Journal. Federal Trade Communications Website. Available at: https://www.ftc.gov/public-statements/2015/09/fccs-knowledge-prob lem-how-protect-consumers-online

[28] Skorup, B. (2016), Who needs the FCC, National Affairs, 41. Available at: https://www.nati onalaffairs.com/publications/detail/who-needs-the-fcc

In the light of the FCC's expansive remit and evidence of its status quo orientation, it would seem that the FTC's mandate related to the protection of privacy and promotion of competition offers better prospects related to the regulation of Big Tech (Hoofnagel, Hartzog, & Solove 2019).[29] The FTC's enforcement mandate and oversight of the Children Online Privacy Protection Act (COPPA), for example, resulted in a $170 million fine levied against Google and its subsidiary YouTube in particular in 2019 for collecting personal information from children without the consent of their parents. This fine had to do with the persistent identifiers/cookies that were used to track users of children's channels to deliver targeted ads (Henderson 2019).[30] Although the FTC has in the past not blocked key acquisitions that in hindsight may have resulted in anti-competitive behaviour such as Google's acquisition of the internet advertising server in 2007 for $3.1 billion, it is clear that there is a greater understanding today of the extent that a handful of companies have in the control over both personal and non-personal data and the data-driven knowledge economy. The FTC's planned review by its Technology Enforcement Division established in 2019 of all the major acquisitions and mergers by Alphabet, Apple, Amazon, Facebook, and Microsoft over a 30-year period will include 767 acquisitions (235—Google), (125—Apple), (101—Amazon), (82—Facebook), and (237—Microsoft) including the acquisition of WhatsApp by Facebook ($19 billion) and LinkedIn by Microsoft ($26.2 billion) (Dan 2020).[31] The aims of the investigation including the assessment of anti-competitive conduct and the robustness of extant anti-competition law. The five companies are expected to furnish documents on corporate acquisition strategies, board appointments, hiring processes, product development and pricing, the

---

[29] Hoofnagle, C. J., Hartzog, W., & Solove, D. J. (2019), The FTC can rise to the privacy challenge, but not without help from Congress, Brookings, 8 August. Available at: https://www.brookings.edu/blog/techtank/2019/08/08/the-ftc-can-rise-to-the-privacy-challenge-but-not-without-help-from-congress/

[30] Henderson. J. G. (2019), Google and YouTube will pay record $170 million for alleged violations of children's privacy law, Federal Trade Commission, 5 September. Available at: https://www.ftc.gov/news-events/press-releases/2019/09/google-youtube-will-pay-record-170-million-alleged-violations

[31] Dan, E. (2020), What does the FTC intend to do about the big tech acquisitions of the last thirty years?, Forbes, 13 February. Available at: https://www.forbes.com/sites/enriquedans/2020/02/13/what-does-the-ftc-intend-to-do-about-the-big-tech-acquisitions-of-the-last-30years/?sh=55dcf65230ab

integration of their acquired assets with the parent company, and treatment of acquired data (Wise, Pinegar, & Walser 2020).[32]

In the context of a new administration taking over in Washington in 2021, it will be interesting to see to what extent there will be attempts to regulate Big Tech in the domestic market and simultaneously work towards cooperation with the EU and other jurisdictions on matters related to privacy, competition, taxation of the digital companies among other issues. If, as Lawrence Lessig famously said code is law, who should have the authority to script that code and thereby, shape individual behaviour? Governments or Big Tech? And what role should civil society play in the shaping of such code? (see Platform Accountability 2020).[33] There are major issues at stake in the context of the global digital economy and trade in digital goods and services including the need for basic agreements on privacy as a fundamental right as is the norm in the EU rather than as an issue linked to consumer protection. Issues such as data portability and the right to be forgotten are founded on the need for competition and consumer choice. The global experience of hate speech and misinformation via platforms owned by Big Tech companies has heightened awareness of the limits of absolute freedom of expression although there is a task ahead for larger learning of the rights and responsibilities of free speech and for appropriate legislations. But it remains to be seen whether Section 230 of the Communications Decency Act, 1996, which protects social media sites such as Facebook from being sued for content hosted by third parties, will be repealed by the Biden administration given their shared liberal values and close connections with Silicon Valley and whether Lina Khan who has had a strong anti-trust record will recommend the reform of platforms in her role as the Commissioner of the FTC (Oremus 2021).[34]

[32] Wise, M., Pinegar, N., & Walser, M. (2020), US Merger control in the technology sector, Law Reviews, September 2020. Available at: https://thelawreviews.co.uk/edition/the-merger-cont rol-review-edition-11/1229561/us-merger-control-in-the-technology-sector

[33] Platform Accountability: Global challenges and opportunities (1–30), Cedar Partners & Associates, 20 November 2020. Available at: https://drive.google.com/file/d/1S4MBS8VmKC iqqBXLdANiF4ijaqfvq-mY/view

[34] Oremus, W. (2021), Biden has a chance to reshape tech. Will he?, OneZero, 24 January. Available at: https://onezero.medium.com/how-biden-could-reshape-the-internet-134a7 726df83

5

The Contrary Compulsions of a Surveillance State

The Equalisation Levy in India

Having written a number of critical political economy-inspired texts on the media and digital India, I am of the opinion that the regulatory role of the Indian state has inherently been ambivalent. By ambivalent, I mean a State that is caught between the need to harmonize its laws with global laws and that is involved in negotiating internal and external pressures to both privatize and protect, resulting in an approach to regulation that can be both 'progressive' such as the Equalisation Levy and 'regressive' such as the Personal Data Protection Bill (PDPB, 2019).[1] This Bill, that was withdrawn in August 2022, was based on the GDPR although it has raised 'concerns regarding the non-effectiveness of consent process in light of technological developments, lack of effective stakeholder participation in the top-down approach of DPAI (Data Protection Authority of India) and the non-serious approach towards data protection by governmental institutions ... ' (Deva Prasad & Menon 2020:19).[2] Another broadly progressive measure is the launch of a super platform, the Open Network for Digital Commerce (ONDC) in April 2022. This can be viewed as an indirect attempt to regulate the reach and power of e-commerce giants such as Amazon and Walmart. The ONDC was intentionally established by the Department for Promotion of Industry and Internal Trade (DPIIT) under the Ministry of Commerce and Industry to create a level playing

---

[1] The Personal Data Protection Bill, 2018. Available at: https://www.meity.gov.in/writereadd ata/files/Personal_Data_Protection_Bill,2018.pdf

[2] Deva Prasad, M. & Menon, S. C. (2020), The Personal Data Protection Bill (2018): India's regulatory journey towards a comprehensive data protection law (1–19), International Journal of Law and Information Technology, 28.

Platform Regulation. Pradip Ninan Thomas, Oxford University Press. © Pradip Ninan Thomas 2023.
DOI: 10.1093/oso/9780192887962.003.0005

field in e-commerce and offer small and micro-retailers opportunities to sell products and services online and compete with established vendors.[3] While it is certainly the case that the contemporary State in India is committed to regulating the digital economy and in particular data controllers such as Facebook, Google, and Amazon it would seem that it is less committed to protecting the privacy of its key data generators—the average Indian citizen. So while the Personal Data Protection Bill for the purposes of government publicity was touted as a bill modelled on the GDPR, the two are fundamentally different precisely for the reason that the digital rights of citizens in the GDPR are considered a substantive human right while in the case of India this has been interpreted as a constitutional right that is located nevertheless within considerable government oversight (Chatterjee 2018).[4] Moreover, unlike the GDPR, which supports the protection of data mobility within the EU while restricting cross-border transfers of data, in the case of this Bill in India, sensitive and critical data (financial, health, biometric) is expected to be stored locally while other forms of personal data can be freely transported (Parkin 2019).[5]

While successive post-Independent governments in India have consistently exerted control over its citizens, the incumbent government, with its overtly Hindu nationalistic predilections and anti-minority stance, presides over a number of surveillance instruments, some of which it inherited. These include the following monitoring systems and databases: the Unique Identity Authority of India (UIDAI), State Resident Data Hubs, Central Monitoring System, Lawful Intercept and Monitoring Project, National Intelligence Grid, Network Traffic Analysis, Crime and Criminal Tracking Network, Interoperable Criminal Justice System, National Cybercrime Coordination Centre, Cities Database, DBA Databanks, National Social Registry, National Digital Health

---

[3] Barik, S. (2022) Explained: Government's open network for digital commerce, and what Microsoft joining it means, The Indian Express, Aug.10. Available at: https://indianexpress.com/article/explained/explained-sci-tech/microsoft-open-network-digital-commerce-india-e-commerce-8082037/

[4] Chatterjee, S. (2018), Is data privacy a fundamental right in India? An analysis and recommendations from policy and legal perspectives (170–190), International Journal of Law and Management, 61 (1).

[5] Parkin, B. (2019), India proposes first major data protection law, The Financial Times, 11 December. Available at: https://www.ft.com/content/df6fd8d4-1bf1-11ea-9186-7348c2f183af

Mission and National Health Stack, National Register of Citizens, National Population Register, National Technical Research Organisation, Public Credit Registry, sector-specific data projects such as *Vahan* (Vehicles) and *Sarathi* (driving licences), *Aarogya Setu* (patient tracking) among a plethora of other initiatives. These surveillance mechanisms are backed by law, in particular, the amended IT Act (2008), itself based on a colonial-era legislation, the Indian Telegraph Act (1885). Section 66A of this act was arbitrarily used by the government to stifle free speech until it was struck down by the Supreme Court in 2015. It is this arbitrary misuse of law and the lack of transparency that is highlighted in this comment by Litton (2015:801)[6] on surveillance in India: 'Nine government entities (The Central Bureau of Investigation ("CBI"), the Narcotics Control Bureau ("NCB"), DRI, National Intelligence Agency, CBDT (tax authority), Military Intelligence of Assam and JK and Home Ministry)—including two spy agencies (Intelligence Bureau & Research Analysis Wing)—will have virtually unfettered access to the sensitive personal information collected through CMS with no court order required to monitor targets, no parliamentary oversight, and no formal privacy regime in place to protect individuals from government intrusion. One expert group created to outline principles for an Indian privacy law described CMS as 'an unclear regulatory regime that is non-transparent, prone to misuse, and that does not provide remedy for aggrieved individuals'. Another example of surveillance in a Covid environment is the Integrated Disease Surveillance Project and the Aarogya Setu mobile app used to identify clusters and hotspots that are both operating in an environment where privacy remains a contested and unresolved issue (Patnaik & Pratap 2020).[7]

I would like to in this chapter deal with two regulatory responses from the government of India to the challenges posed by the data economy—the Personal Data Protection Bill and the Equalisation Levy.

---

[6] Litton, A. (2015), The State of Surveillance in India: The Central Monitoring System's chilling effect on self-expression (799–821), Washington University Global Studies Law Review, 14 (4). Available at:https://openscholarship.wustl.edu/cgi/viewcontent.cgi?article=1556&context=law_globalstudies

[7] Patnaik, A. & Pratap, N. (2020), India needs a surveillance law that goes beyond personal data protection, The Wire, 9 June. Available at: https://thewire.in/tech/india-surveillance-law-personal-data-protection

The Personal Data Protection Bill, 2019

In early 2018, following some anxiety and confusion over whether the Aadhar, Unique Identity card will be a requisite means to authenticate data transactions in India, I decided as an Overseas Citizen of India (OCI) to enrol in this scheme. Since most people in Chennai had already enrolled, there were few enrolment centres that were open although there was a facility for mobile, home enrolment. My wife and I decided on the latter—and early one morning, a gentleman appeared with a fingerprint and iris scanner and promptly and rather painlessly gave us both our 12-digit identities while enrolling us into this nationwide database. Since no questions were asked, it was assumed that as brown-skinned Tamil speakers, we were legitimate and resident at the address that we were at. While the information that we furnished for the 'fields' was verifiable and authentic and all documents used for the authentication process from passports and PAN cards, etc. were legal, there was a certain laxity about the process that could have been exploited by a person with nefarious intent to generate a unique identity for scamming welfare payments and/or for criminal ends. While not on the same level of absurdity as reports of dogs and cows being enrolled in the scheme—it nevertheless did demonstrate negligence in the collection of personal data and 'authentication' in what is a nation scale project. This attitude towards the collection of private data is reflected in the Indian government's attitude towards privacy that has been of secondary concern in a country where successive governments have had a patrician and overwhelmingly top–down approach to dealing with citizen data. To some extent, the citizens' movement related to the Right to Information that began in the late 1980s and that eventually led to the Right to Information Act in 2005 was the first occasion when transparency and accountability became the litmus test for public–private transactions. This national movement made a clear case for the sanctity and non-transmutability of personal data, respect for its inviolability, and democracy as rule by people. The continuing attempts to dilute citizen rights to know and expand state 'exemptions' are wholly in character with the present government.

In a study of data protection in Indian law by Graham Greenleaf (2011),[8] the patchy, disparate nature of privacy law in India is discussed

---

[8] Greenleaf, G. (2011), Promises and illusions of data protection in Indian law (47–69), International Data Privacy Law, 1 (1).

against India's ambitions to become one of the world's premier personal data processing centres. The massive investments in business process outsourcing, medical transcription services, and international banking services were, however, not backed up by any regulatory safeguards—and while there have been occasional leaks of personal data, global scrutiny of the risk factor in personal data processing in India has not been of major concern. A decade later and in the context of social media and India becoming a destination for commercial content moderation, the protection of private data has become a primary concern, although arguably there has been a lack of urgency in the State's response. Greenleaf's survey looks at data protection in the Information Technology Act 2000 (amended 2008), Credit Information Companies (Regulation) Act 2005, The Right to Information Act (2005), the Protection of Human Rights Act (19930), The National Consumer Disputes Redressal Commission, the Indian Contract Act, 1872 and data protection under the industry body the National Association of Software and Services Companies (NASSCOM) and its self-regulatory body the Data Security Council of India (51–55). It is interesting that out of these Acts, it is the Credit Information Companies Regulation Act that has been singled out for its 'comprehensive data protection standards' (52) although even in this case, personal data has not been defined and overall, the gap between privacy principles and practice is extensively leading to Greenleaf concluding that 'at present India does not provide significant protection to personal data in relation to all or most of the common privacy principles, in any sector, to meet any international standard' (68). There are, as Abraham and Hickok (2012:303)[9] note in an article on government access to private sector data in India, 50 policies including 'statutes, rules, regulations and executive orders' that implicitly or explicitly protect privacy.

It is fascinating that in the context of the many rulings on privacy—that the Constitution has appeared as a clear winner, the absolute final source for interpretations, re-interpretations, derivative rulings, and substantive understandings of what remains a contentious right. While there have been numerous judgements on privacy in post-Independent India, a key ruling is the verdict in Justice K S Puttaswamy vs. Union of India

[9] Abraham, S. & Hickok, E. (2012), Government access to private-sector data in India (302–315), International Data Privacy Law, 2 (4).

(2017) on Aadhar verifications and State over-reach and citizen rights to privacy. In a historic judgement delivered on 27 August, nine Supreme Court judges averred that while privacy was not an absolute right that it was implicit in Part III of the Constitution of India as a fundamental right highlighted in rights relating to equality (Articles 14 to 18); freedom of speech and expression (Article 19(1)(a)); freedom of movement (Article 19(1)(d)); protection of life and personal liberty (Article 21) (Bhandari et al. 2017).[10] While this judgement affirms privacy as a Constitutional right, it remains less categorical on the inviolability of this right and is in favour of a case-to-case approach to dealing with it. This ruling led to the Ministry of Information and Technology constituting an expert committee under Justice Srikrishna to look into a data protection framework. Their 213-page report—A Free and Fair Digital Economy Protecting Privacy Empowering Indians, in the wake of the Puttaswamy judgement, (https://www.meity.gov.in/writereaddata/files/Data_Protection_Committee_Report.pdf) is benchmarked against international instruments, particularly the EU's GDPR. This report became the basis for the Personal Data Protection Bill that was introduced to Parliament on 11 December 2019 and subsequently referred to a Joint Parliamentary Committee under the chairpersonship of Meenakshi Lekhi from the ruling BJP. Their report was scheduled to be submitted to Parliament during the Winter Session 2020 although the Bill itself was withdrawn by the government in August 2022.

The Bill's preamble outlines the objectives of data protection:

'To protect the autonomy of individuals in relation with their personal data, to specify where the flow and usage of personal data is appropriate, to create a relationship of trust between persons and entities processing their personal data, to specify the rights of individuals whose personal data are processed, to create a framework for implementing organisational and technical measures in processing personal data, to lay down norms for cross-border transfer of personal data, to ensure the accountability of entities processing personal data, to provide

[10] Bhandari, V., Kak, A., Parsheera, S., & Rahman, F. (2017), An analysis of Puttaswamy: The Supreme Court's privacy verdict, Indrastra, 18 November. Available at: https://medium.com/indrastra/an-analysis-of-puttaswamy-the-supreme-courts-privacy-verdict-53d97d0b3fc6

remedies for unauthorised and harmful processing, and to establish a Data Protection Authority for overseeing processing activities' (6, the Personal Data Protection Bill, 2018).[11] While these objectives are laudable including the affirmation that the right to privacy is a fundamental right and that collective cultures and enabling environments are required to further this right, in the context of a control state and the capture of key institutions including the Supreme Court by the incumbent government, there are questions related to the autonomy of judicial interpretations when it comes to privacy and whether or not, in the last instance, personal privacy rights based on the principles of proportionality, justness and fairness will trump the over-reach of the State. Basu & Sherman's (2020)[12] critique of the Bill highlights the issue with State over reach—'... the text of the bill largely appears to be a crude amalgamation of provisions in the GDPR with authoritarian leanings. In the Indian bill, these include the enabling framework for government surveillance in the bill, which undoubtedly entrenches government power to undermine citizen privacy. Further, the blurring of the distinctions between non-personal data and personal data remain is concerning. The bill ultimately dilutes protections on individual data rights by enabling the government to access anything it feels would fit within the laid-out categories of exemptions'. In other words, the Bill does fall short on significant support for data justice while offering substantive possibilities for the State to data discriminate (see Taylor 2017).[13] A response from the Centre for Internet Society, a Bengaluru-based thinktank on to the Bill points out some of its flaws—including shortcomings in its privacy by design approach, its use of Dark Patterns in its interface that can be deceptive and trick users into compliance, the limited number of organisations that fall under the nomenclature Social Media Intermediaries or Fiduciaries that excludes e-commerce firms, search engines and email services along with issues related to 'consent' in the context of access in Indian languages, in plain language and the

---

[11] PDPB, ibid.

[12] Basu, A. & Sherman, J. (2020), Key global takeaways from India's revised personal Data Protection Bill, Lawfare, 23 January. Available at: https://www.lawfareblog.com/key-global-takeaways-indias-revised-personal-data-protection-bill

[13] Taylor, L. (2017), What is data justice? The case for connecting digital rights and freedoms globally (1–14), Big Data & Society, July–December.

availability of notices (Naidu, et al. 2020).[14] So is the PDPB a missed opportunity? An optimist would argue that the content of the Bill based on benchmarking against the GDPR offers the very first opportunity for the state to harmonise its privacy laws with 'progressive' thinking such as the 'right to forget', 'data portability', the responsibilities of 'data inter-mediaries' and the like. The pessimist though, would point to the fact that State oversight is built into the Bill further strengthening the role of the state in the surveillance and disciplining of its citizens. While there are real threats to the integrity of the nation from terrorists that should not be discounted, the State also has cultivated many imagined threats linked to citizen transactions of data. Data privacy laws remain vague, opaque and open to a variety of interpretations. This may be deliberate given that the more explicit policy is, the more pressure for the govern-ment to abide by the principles related to citizens and their control over their data.

The rather abrupt withdrawal of the Personal Data Protection Bill on August 4, 2022 by the government of India, ostensibly in response to negative feedback from privacy advocates and data corporates along with the evidence of the lack of clarity with respect to government over-sight and terms of access to personal data, suggests the persistence of the control paradigm that has been favoured by successive Indian govern-ments.[15] Eighty one amendments were suggested for a Bill that consisted of ninety one sections.[16] An alternate plan is now being considered for a 'comprehensive legal framework' that will be used to establish separate laws for data privacy and other issues that require regulatory oversight. So the matter of who has control over personal data in India remains, at least for the moment, in regulatory limbo.

---

[14] Naidu, S., Seshadri., A., Mohandas, S., & Bidare, P. M. (2020), The PDP Bill 2019 through the lens of privacy by design, Centre for Internet & Society, 12 November. Available at: file:///Users/uqpthom4/Downloads/PDP%20Bill%20Design%20Analysis%20V2.pdf

[15] Singh, M. (2022) India withdraws personal data bill that alarmed tech giants, TC, Aug. 4. Available at: https://techcrunch.com/2022/08/03/india-government-to-withdraw-personal-data-protection-bill/

[16] Y. Sameer & Singh, K. D. (2022), India withdraws a proposed law on data protection, The New York Times, August 4. Available at: https://www.nytimes.com/2022/08/04/business/india-data-privacy.html

Aadhar and Privacy

Aadhar offers a salient case study of how privacy or the lack of it affects those on welfare disproportionality to those who are more comfortably off in India. In other words, one's class and caste positions along with vocation can affect one's use of Aadhar. Biometrics-based ID systems based on fingerprints and iris scans inherently discriminate because such systems cannot capture unique data from those who live on the basis of manual labour and whose bodies are shaped by precarity. These are also the very communities who are in the deep end of the digital divide and who do not have easy access to smartphones, lack the required literacies, and do not have support systems. Yadav (2016),[17] in an article on the roll-out of Aadhar in districts in the Western Indian state of Rajasthan, highlights the fact that in the district of Ajmer alone (there are 33 districts in Rajasthan), at each of the 1,145 ration shops, the oldest and poorest faced issues with the non-authentication of their biometric data while those belonging to the same cohort also did not have access to smartphones. Recognition, in particular the lack of it, compromises human dignity and remains a challenge along with the more widespread critique of State surveillance (Singh 2019).[18] I could, as an entitled OCI, explore Aadhar as a process, safe in the conviction that this would not compromise my right to survive but useful as an ID when next boarding a domestic flight in India. For the less fortunate though, Aadhar could be the difference literally between life and death. While privacy may not be a major concern for those who depend on the welfare system for entitlements such as food grains and essential oils for survival, this may, in the context of moves by the incumbent government to corral, de-limit, and disenfranchise citizens who belong to minority groups, scheduled castes, and indigenous communities, impact on their Constitutional right to equality as Citizens of India. If the Aadhar identity is linked to voter identity, this could, in a polarized communal climate, be used to disenfranchise minority communities who are

---

[17] Yadav, A. (2016), In Rajasthan there is 'unrest at the ration shop' because of error-ridden Aadhar, 2 April, Scroll.in. Available at: https://scroll.in/article/805909/in-rajasthan-there-is-unrest-at-the-ration-shop-because-of-error-ridden-aadhaar

[18] Singh, P. (2019), Aadhar and data privacy: Biometric identification and anxieties of recognition in India (1–16), Information, Communication & Society, https://doi.org/10.1080/1369118X.2019.1668459

already part of the precariat and on the edge. Moreover, in keeping with the control State, the Aadhar (Targeted Delivery of Financial and Other Subsidies, Benefits and Services) Bill, 2016 is silent on issues related to privacy and delegates the redressal of grievances to the very body that administers the initiative—the UIDAI under Clause 23 (2)(s) of the Bill (Thikkavarapu 2016).[19] This rather deliberately cavalier (see Ramnath & Assisi 2018)[20] approach to the avoidance of scrutiny that is based on all power invested in a body that administers, maintains, rules over and against Indian citizens and that is not beholden to any other authority reflects the exercise of overwhelming power. Khera (2019:72–85)[21] in an edited volume on Aadhar offers five privacy concerns related to this initiative linked to the compromise of personal integrity, bodily integrity, data integrity, its integration with centralized and inter-linked surveillance databases, and the lack of clarity on whether personal data in the personal data economy will become monetized.

The government's overall approach to privacy, highlighted in its formulation of the PDB, does, however, seem to be at odds with the spirit or lack of it in the Draft India Data Accessibility and Use Policy (2022)[22] issued by the Ministry of Electronics and Information Technology (Meity). The intent of this document is to explore the commercial use and exploitation of public data sets. With little privacy safeguards in place, this policy that invokes the benefits of open data and intra-government sharing of such data, explicitly highlights the use of high-value data sets and the licensing of data sets to commercial vendors. Based on little consultation with stakeholders, this policy remains ambivalent about privacy although it is forthright about the role played by the private sector in data value creation that would enable the shaping and making of a US$5 trillion digital economy. It is interesting that the principle of open data and the need for it to be commercially exploited is fulsomely dealt with in

[19] Thikkavarupu, P. R. (2016), The Aadhar Bill is yet another legislation that leaves too much power with the government at the centre, The Caravan, 16 March. Available at: https://caravanmagazine.in/vantage/aadhaar-bill-another-legistlation-leaves-power-centre

[20] Ramnath, N. S. & Assisi, C. (2018), The Aadhar Effect: Why the world's largest identity project matters, Oxford University Press, New Delhi.

[21] Khera, R. (2018), Aadhar and privacy (72–85) in Khera, R. (Ed.), Dissent on Aadhar: Big Data Meets Big Brother, Orient BlackSwan, Hyderabad.

[22] India Data Accessibility and Use Policy (2022), Ministry of Electronics and Information Technology, February. Available at: https://www.medianama.com/wp-content/uploads/2022/02/Draft-India-Data-Accessibility-and-Use-Policy.pdf

the chapter 'Data "Of the people, by the people, for the People"' (78–97), Chapter 4, Economic Survey 2018–2019[23] although the optimism about existing data privacy measures seems misplaced—'there is tremendous scope for the private sector to benefit from the data and they should be allowed to do so, at a charge. Fortunately, stringent technological mechanisms exist to safeguard data privacy and confidentiality even while allowing the private sector to benefit from the data' (94).

The Equalisation Levy

The moves to regulate key players in the big data economy have gathered steam over the last decade. In countries around the world, the regulation of big data companies has become the focal point for multiple deliberations and actions. In the United States alone, after a regulatory hiatus that enabled big data companies to become virtual monopolies, there are moves by the Federal Trade Commission to investigate below-the-radar acquisitions and acqui-hires by Alphabet, Amazon, Apple, Facebook, and Microsoft, anti-trust actions by the House Judiciary Committee, The Justice Department's anti-trust actions against Google, at the level of individual states such as in Texas—49 offices involved in an anti-trust investigation of Google among numerous investigations into big data companies (Brandom 2020).[24] The Deceptive Experiences to Online Users Experiences (DETOUR) Act, 2019 is a bipartisan act specifically meant to prohibit online platforms from 'using deceptive user interfaces, known as "dark patterns" to trick consumers into handing over their personal data' (US Senator Deb Fisher 2019).[25] The EU has been the most dogged in trying to create a level playing field for big data companies

---

[23] Data 'Of the people, by the people, for the People' (78–97), Chapter 4, Economic Survey 2018–2019, Ministry of Finance, https://www.indiabudget.gov.in/budget2019-20/economicsurvey/doc/vol1chapter/echap04_vol1.pdf

[24] Brandom, R. (2020), The regulatory fights facing every major tech company: Facebook, Google, and Amazon versus the world, The Verge, 3 March. Available at: https://www.theverge.com/2020/3/3/21152774/big-tech-regulation-antitrust-ftc-facebook-google-amazon-apple-youtube

[25] Senators introduce bipartisan legislation to ban manipulative 'dark patterns' (2019). Press Release, United States Senator Deb Fisher for Nebraska, 9 April. Available at: https://www.fischer.senate.gov/public/index.cfm/2019/4/senators-introduce-bipartisan-legislation-to-ban-manipulative-dark-patterns

with the Digital Services Act, 2020 being the clearest attempt to manage content mined by big data and strengthen competition (Amaro 2020).[26] The EU's Competition Commissioner Margrethe Vestager has among a number of investigations into the practices of big data companies in the EU, launched a major probe into how Amazon uses its market dominance to function both as a marketplace for merchants and as a rival seller of all manner of goods and services (Yun Chee 2020).[27] The EU has fined Big Tech companies in billions of dollars and, over the last decade, a range of big data company practices have invited sanctions and fines, there have been investigations related to data storage and privacy, labour infringements by companies such as Uber, National Security breaches, and Tax infringements.

The Equalisation Levy is a specific response to tax infringements by Big Tech companies, and is an attempt to level the playing field and ensure that taxes are paid within the jurisdictions that these companies operate in. It is a response to the fact that governments have lost revenues, local companies have not been able to compete and tax systems compromised because of their inability to tax major players in the digital economy. Giant online intermediaries, straddling the globe, with their assets mainly tied up in 'intangibles', big data, and personal data and involved in creating value from multiple externalities and billions of transactions—are so very different from regular brick and mortar companies that are taxed on the value that they generate through manufacturing, providing services, etc. based on physical presence. The UK-based aid agency Action Aid has estimated that the tech giants responsible for the largest tax gaps (US$2.8 billion) operate in India, Brazil, Indonesia, Nigeria, and Bangladesh, and include Google, Facebook, and Microsoft (Facebook, Google and Microsoft 2020).[28] In Europe, France has been vigorously pursuing back taxes and earned €106 million from Facebook for 2009–2019.

---

[26] Amaro, S. (2020), The EU is about to announce new rules for big tech—and there's not much they can do about it, CNBC, 5 November. Available at: https://www.cnbc.com/2020/11/05/digital-services-act-how-the-eu-is-going-after-big-tech.html

[27] Yun Chee, F. (2020), Europe charges Amazon with using dominance and data to squeeze rivals, Sydney Moring Herald, 11 November. Available at: https://www.smh.com.au/business/companies/europe-charges-amazon-with-using-dominance-and-data-to-squeeze-rivals-20201111-p56dfg.html

[28] Facebook, Google and Microsoft 'avoiding $3bn in tax I poorer nations' (2020), BBC, 26 October. Available at: https://www.bbc.com/news/business-54691572

Facebook paid 50% more in tax to authorities in France than it paid in 2019—€8.46 million in 2020 (Facebook Agrees to Pay 2020).[29] In 2019, Google paid €1 billion in unpaid taxes to France after it was found that while it operates in the EU, it reports its sales in Ireland, a country whose tax threshold of 12.5% is lower than in other EU countries. These are but the tip of the iceberg and the basis for major confrontations with authorities in the United States and retaliatory actions by them against the EU and France. The OECD that has played a central role in advancing a unified global digital tax policy through its Base Erosion and Profit Sharing (BEPS)[30] initiative, through its Forum on Harmful Tax Practices (FHTP) identified 12 jurisdictions where no or nominal taxes are paid including no or nominal corporate taxes. These include Anguilla, the Bahamas, Bahrain, Barbados, Bermuda, British Virgin Islands, Cayman Islands, Guernsey, Isle of Man, Jersey, Turks and Caicos Islands, and the United Arab Emirates. This was part of a larger investigation that dealt with 287 preferential tax regimes (Asen 2019).[31]

In 2012, G20 finance ministers at the leader's summit in Los Cabos, Mexico, requested the OECD to develop an action plan linked to dealing with the tax challenges faced in a digital economy. In 2013, released its 15-point Action Plan on Base Erosion and Profit Shifting (Action Plan 2013)[32] followed by an inclusive framework for BEPS implementation (2015). The creation of a comprehensive framework for the taxation of digital transactions is complicated precisely for the reason that every user is a creator of value. In other words, it is this data input from users that create value for the platforms. They link this value to advertisers who create better-targeted advertising for the sellers of goods and services online. As a non-rivalrous good, the cost of production and distribution of copies of any given unit of a digital good is zero. These products are highly mobile and the business model is based on the management and

---

[29] Facebook agrees to pay France €106m in back taxes, BBC, 24 August. Available at: https://www.bbc.com/news/business-53894959

[30] BEPS inclusive framework on base erosion and profit sharing. Available at: https://www.oecd.org/tax/beps/about/

[31] Asen, E. (2019), OECD tackling harmful tax practices, Tax Foundation, 18 September. Available at: https://taxfoundation.org/oecd-harmful-tax-practices-base-erosion-profit-shifting/

[32] Action on base erosion and profit shifting (2013), OECD. Available at: https://read.oecd-ilibrary.org/taxation/action-plan-on-base-erosion-and-profit-shifting_9789264202719-en#page1

monetization of intergroup network effects involving customers, sellers, and advertisers (Kind & Koethenbuerger 2018).[33] OECD member countries have come up with three solutions (1) a taxation regime based on the taxation of income attributable to users involved in the creation of significant data and brand value, (2) marketing of intangibles—the taxation of value created through trade-in and the multi-sectoral exploitation of data generated through marketing goods and services, and assets based on algorithms and artificial intelligence, and (3) taxation that is based on the significant economic presence of firms that are not located in the jurisdictions that they operate in. While it is less of a challenge to levy taxes on companies whose key revenues are generated through digital advertising revenues such as Facebook and Google, apprehending and interpreting the creation of digital value remains a major challenge. A chapter from a volume published by the OECD highlights the challenge of understanding and calculating the creation of value by users 'of a participative networked platform contribute user-created content, with the result that the value of the platform to existing users is enhanced as new users join and contribute. In most cases, the users are not directly remunerated for the content they contribute, although the business may monetise that content via advertising revenues (as described in relation to multi-sided business models below), subscription sales, or licensing of content to third parties' (Broader Tax Challenges: 2014:128).[34] How to understand both incremental and aggregated value remains a challenge. In other words, the need to align income taxation with value creation, a stated goal of the OECD, is a principle that most jurisdictions agree to although it is hard to implement (Grinberg, 2019:89).[35] The calculation of tax on digital intangibles does seem like a Herculean task in the transactional economy. In 2019, OECD released its Inclusive Framework on

---

[33] Kind, H. J. & Kothenbuerger, M. (2018), Taxation in digital media markets (22–39), Journal of Public Economic Theory, 20.

[34] Broader tax challenges raised by the digital economy (123–139) in Addressing the Tax Challenges of the Digital Economy (2014), OECD Publishing, Paris. Available at: https://www. oecd-ilibrary.org/docserver/9789264218789-10-en.pdf?expires=1574653950&id=id&accn ame=guest&checksum=2AE4B3E2772AF2873DFBB51A8DEB653D

[35] Grinberg, I (2019), International taxation in an era of digital disruption: Analysing the current debate (85–118), Taxes, March. Available at: https://scholarship.law.georgetown.edu/cgi/ viewcontent.cgi?article=3163&context=facpub

BEPS that enables interested countries with a tool kit to detect and resolve the issue of tax avoidance—thus contributing to a more equitable global tax regime.

In the context of major delays over a global digital tax policy, the Government of India, also a member of the OECD, decided in 2016 to unilaterally create its own digital tax regime. In June 2016, the government of India decided that all online information and database access or retrieval (OIDAR) services would invite a levy. Seven services were included under the OIDAR including advertising on the internet, cloud services, and online gaming although the levy was primarily focussed on companies involved in advertising on the internet. Chapter VIII of the Finance Act was amended to include a 6% levy on business-to-business transactions linked to digital advertising carried out by non-Indian companies that had a substantial economic presence in India but whose headquarters were elsewhere. The B2B business includes 'Social media companies, Internet search companies, Digital media, websites App developers and App marketplace (Android)' who 'derive significant revenues from digital advertising on their websites, mobile sites, mobile apps, etc who derive advertisement revenues from hosting ad banners, related content ads (sidebar ads), target ads, ad word searches, pre-roll videos based on user preferences, etc. E-commerce companies/online travel/hotel sites though not primarily dependent also rely partly on revenues from product advertising, promotions, etc' (Deloitte, Equalisation Levy, 2016:2).[36] The levy is withheld by the Indian company that uses these services and that is paid to the government as a form of 'presumptive' tax, an indirect means to ascertain tax liability that is a solution to a hard to tax issue. Fundamentally, this tax is a solution to the issue of double non-taxation and the jurisdiction of a company with no permanent establishment in India but with substantive economic presence. In 2018 the concept of 'substantive economic presence' was added to considerations under the Equalisation Levy, followed in April 2020 by the expansion of the levy to cover e-commerce operators involved in the sale of digital goods and services and amounting to 2% of revenues earned

---

[36] Deloitte (2016), Equalisation Levy, 2016: Is it equitable? (1–8), Deloitte, June. Available at: www.deloitte.com/in

by non-resident e-commerce operators (Jain, Nagappan, & Aggarwala 2020)[37] whose annual revenues are in excess of \$237,000. Since 2016, the government of India has earned Rs. 4,000 crores (\$541 million) from this levy (Mondal 2020).[38] Rather predictably, the USTR has launched a Section 301 investigation as it has in other jurisdictions where a 'Google tax 'has been applied to companies that are of US origin and that are headquartered in the United States. Austria has levied a 5% tax that covers online advertising, France has levied 3% on advertising services while Turkey has levied a 7.5% on online services including advertisements and sales of content (Shashidhar & Parpiani 2020).[39] The United States has already raised issues with India's effort to restrict cross-border data flows and mandate data localization. However, in the light of the fact that companies such as Apple have decided to pass on the price of this levy to cover all Apple products sold in India (Jalan 2020),[40] a Section 301 investigation on this 2% tax would seem pointless although there probably are other trade-related issues that have motivated this investigation.

The Personal Data Protection Bill and the Equalisation Levy are two responses from a state that tends to veer towards surveillance when it comes to dealing with personal data while attempting to bring foreign-owned digital data companies within the ambit of Indian taxation law. This imbalance is reflected in yet another bill that is on the anvil—the DNA Technology (Use and Application) Bill, which is ostensibly meant to legalize DNA databanks of genetic material from those the state considers to be of a criminal bent and anti-national. In the absence of strong privacy laws in India and in the context of numerous cases of ordinary

---

[37] Jain, P., Nagappna, M., & Agarwalla, I. (2020), India Equalisation Levy expanded—a surprise move! International Forum: Bloomberg Tax, 29 May. Available at: https://news.bloom bergtax.com/daily-tax-report-international/india-equalization-levy-expanded-a-surprise-move-part-1

[38] Mondal, R. (2020), India collected Rs. 4,000 crore 'Google Tax' since 2016; Rs 1,000 crore in FY20. Business Today, 20 July. Available at: https://www.businesstoday.in/current/economy-politics/india-collected-rs-4000-crore-google-tax-since-2016-rs-1100-crore-in-fy20/story/410545.html

[39] Shashidhar, K. J. & Parpriani, K. (2020), Understanding America's response to India's Equalisation Levy, Observer Research Foundation. 18 July. Available at: https://www.orfonline.org/expert-speak/understanding-america-response-india-equalisation-levy/

[40] Jalan, T. (2020), Apple app store prices to increase reflecting 2% equalisation levy, Medianama, 27 October. Available at: https://www.medianama.com/2020/10/223-apple-store-equalisation-levy-india/

citizens being charged under the colonial-era sedition law, the potential for the misuse of this law is immense. Nayantara Ranganathan (2021)[41] writing in the *Hindustan Times* highlights some of the flaws with this Bill. 'Experts believe that the bill leaves ample room for misuse, and that its consent provisions are not strong. A more fundamental concern is that DNA technology for identification derives from antiquated and discredited methods. Scientists confirm that much of DNA analysis involving statistical modelling algorithms embed judgments of the people behind the creation of these tools. This means that DNA samples collected are used to statistically create composites of "types" of people— racial, ethnic and so on. These methods, in their composition of types, in the inferences drawn, and the mathematical fact of computing averages to arrive at the estimates of types, have the scope for giving a scientific varnish to existing social and cultural bias'. Another attempt to regulate platforms in India is the notification that was sent on 9 November 2020 to the 40+ over-the-top (OTT) platforms such as Netflix, Amazon Prime, and Hotstar (Disney Plus) involved in the provisioning of streaming services informing them that their services will fall under the purview of the Ministry of Information and Broadcasting (MIB) thus creating a level playing field and extending regulation to electronic, print, and digital media (Obhan & Bhalla 2020).[42]

These examples suggest that the Indian government is investing in platform regulation at a variety of levels. Given the ideological trajectory of the incumbent government though both data nationalism and platform nationalism will remain on the agenda and it will not be a surprise if efforts are made to establish Indian platforms on a par with Alibaba and Tencent in China. In 2018, the Indian Yoga guru and entrepreneur Baba Ramdev, launched a messaging app called Kimbho, which in Sanskrit means 'Whatsup'—although it was withdrawn soon after security flaws

[41] Ranganathan, N. (2021), The DNA Bill will cement a disturbing link between tech and policing, The Hindustan Times, 11 February. Available at: https://www.hindustantimes.com/opin ion/the-dna-bill-will-cement-a-disturbing-link-between-tech-and-policing-101612965364 811.html

[42] Obhan, A. & Bhalla, B. (2020), OTT platforms brought under government regulation, Obhan & Associates. Available at: https://www.obhanandassociates.com/blog/ott-platforms-brought-under-government-regulation/?utm_source=Mondaq&utm_medium=syndicat ion&utm_campaign=LinkedIn-integration

were discovered (Bhushan & Aulakh 2018).[43] Key issues in India include the following—how the government will manage competition in the platforms space, whether the privacy of its citizens will be translated into legal provisions and to what extent the control paradigm will determine the trajectory of platform regulation.

[43] Bhushan, R. & Aulakh, G. (2018), Baba Ramdev withdraws messaging app 'Kimbho' after test run, The Economic Times, 1 June. Available at: https://economictimes.indiatimes.com/tech/internet/baba-ramdev-withdraws-messaging-app-kimbho-after-test-run/articleshow/64408760.cms

6

The Australian Consumer and Competition Commission's (ACCC) Digital Platforms Inquiry

With a 25 million population and a risk-averse politics, Australia is not by any means known for its power to shape global regulation or to dictate the terms for the operations of multinational corporations either located in Australia or who have a substantial economic presence. In fact, Australia is a conservative country and an example of this conservatism is the domination of its retail landscape by two monopoly groups, Coles and Woolworths, and its media by two large companies—Rupert Murdoch's News Corp which owns to close to 60% of all newspaper output and Channel 9, a broadcasting company that took over Fairfax, the other major print monopoly in 2017 following the abolishment of cross-media ownership laws that prevented any group that had substantive shares in two media areas (such as TV and print) from investing in a third area (radio) or had a reach/coverage of 75% of the population (McDuling 2018).[1] The Murdoch press has had a consistent reputation for being partisan, supporting the Liberal/National Coalition and being actively involved in engineering political succession both within the Coalition and in the context of national electoral politics. This significant investment in politics and reputation for circulating misinformation on climate change led to calls from a former Labour prime minister Kevin Rudd and former Coalition PM Malcolm Turnbull to institute a Royal Commission into the operations of News Corp in Australia, a process that has begun with a

---

[1] McDuling, J. (2018), What does the Nine-Fairfax merger mean?, Sydney Morning Herald, 7 December. Available at: https://www.smh.com.au/business/companies/what-does-the-nine-fairfax-merger-mean-20181204-p50k1o.html

Platform Regulation. Pradip Ninan Thomas, Oxford University Press. © Pradip Ninan Thomas 2023.
DOI: 10.1093/oso/9780192887962.003.0006

Senate hearing (Kelly 2020).[2] During political negotiations for the repeal of cross-media ownership laws in 2017, support from Independent MPs, in particular Nick Xenophon, was secured in exchange for an enquiry into the digital dominance of platforms in digital advertising and its implications for the future of public interest journalism (Taylor 2018).[3] This was to the advantage of media monopolies including News Corp, that had been involved in lobbying the government to institute an enquiry into digital platforms in Australia. News Corp submitted an 80-page report to the Australian Competition and Consumer Commission (ACCC) that was mandated to establish an enquiry into platforms and their impact on the survival of journalism in Australia in the context of the migration of close to 80% of ad revenues to Google and Facebook (News Corp Submission to ACCC 2019).[4] It was during Malcolm Turnbull's tenure as PM (2015–2018) that processes related to the establishment of a Consumer Data Rights (CDR) protection Bill were initiated. Its origins are linked to an inquiry by the Senate Economics Committee into the anti-consumer practices of the four big banks in Australia—the ANZ, CBA, NAB, and Westpac. This bill that became law in 2019 has been included in Part IVD of the Competition and Consumer Act (2010), the Australian Information Commissioner Act (2010), and the Privacy Act (1988) and is applicable to consumer data protections related to banking, energy, and telecommunications (Treasury Laws Amendment Bill 2019).[5] The CDR when fully operational, will enable data portability rights for users across the banking, energy, and telecommunications sectors.

I begin with this rather involved history of the background to the ACCC's enquiry into digital platforms because it highlights both the real challenges faced by journalism in Australia in the context of their losing

[2] Kelly, C. (2020), Rudd v Murdoch: The ex-PM has won an enquiry into media diversity, but he won't stop there, The New Daily, 14 November. Available at: https://thenewdaily.com.au/news/politics/australian-politics/2020/11/14/kevin-rudd-media-diversity-murdoch/

[3] Taylor, L. (2018), ACCC will have power to grill Google and Facebook on threat to news media, The Guardian, 26 February. Available at: https://www.theguardian.com/media/2018/feb/26/accc-will-have-power-to-grill-google-and-facebook-on-threat-to-news-media

[4] News Corp submission to the Australian Competitions and Consumer Commission, March 1. Available at: https://www.accc.gov.au/system/files/News%20Corp%20Australia%20%28March%202019%29.pdf

[5] Treasury Laws Amendment (Consumer Data Right) Bill, 2019. Parliament of Australia. Available at: https://www.aph.gov.au/Parliamentary_Business/Bills_Legislation/Bills_Search_Results/Result?bId=r6370

out on ad revenues and the less than a compelling case from News Corp in its professed role as a standard bearer of public interest journalism, given global evidence to the contrary as the Leveson enquiry into the phone hacking scandal in the United Kingdom clearly reveals (Freedman 2012).[6] Arguably, the News Corp empire has presided over the demise of quality journalism, be it via its broadcasting outlets such as SkyNews. au and/or Fox News in the United States and print outlets such as *The Australian* and *The Sun* (UK).

The ACCC's Digital Platforms Report

The ACCC's rather voluminous 620-page final report on digital platforms was submitted to the Australian government in June 2019 (Digital Platforms Inquiry 2019).[7] The inquiry focussed on the impact of online search engines in particular Google and social media in particular Facebook, in particular the impact of these platforms on journalism's business models based on advertising and subscriptions. The enquiry looked at three user groups—namely advertisers, media content providers, and consumers and three overlapping issues—data protection, competition, and consumer protection. The final report highlights the fact that both Google and Facebook have substantial market power in their respective markets, in the supply of advertising/display advertising services (Google earned $4.3 billion in 2019 while Facebook earned $0.7 billion) and in their dealings with news media businesses. The report includes 23 recommendations that are directed towards ensuring competition, consumer protection/privacy, education, compensation, platform-neutral media regulation, further investigations into adtech and advertising services, the development of codes of conduct, the establishment of a specialist Digital Platforms Branch within the ACCC among others. Some of these recommendations have since been accepted by the government and the development of codes of conduct for platform-news

---

[6] Freedman, D. (2012), The phone hacking scandal: Implications for regulation (17–20), Television and New Media, 13 (1).

[7] Digital Platforms Inquiry: Final Report (2019), Australian Competition and Consumer's Commission, 26 July. Available at https://www.accc.gov.au/publications/digital-platforms-inquiry-final-report

media compensation negotiations and investigations into the influence of adtech services are ongoing. In January 2020, the ACCC released its interim report on Digital Advertising Services.[8] However, it is the News Media and Digital Media Mandatory Bargaining Code based on amendments to the Competition and Consumer Act (2010) that was introduced to the House of Representatives on 10 December 2020 and that includes a framework for payment bargaining rules for news media, compulsory arbitration rules, minimum standards, and non-discrimination requirements that have become the basis for a public stand-off between Google in particular and the Australian government. The Exposure Draft of this Bill (2020:18)[9] explicitly states that:

'The bargaining news business corporation for a registered news business may notify a responsible digital platform corporation for a digital platform service that it wishes to bargain over one or more specified issues relating to the registered news business' covered news content made available by the digital platform service'

The report also highlights additional responsibilities for platforms that will include their need to comply with:

'minimum standards which require them to provide registered news businesses with advance notification of algorithm changes, provide information about the collection and availability of user data, develop a proposal to recognise original news and give advance notification of changes affecting the display and presentation of news content thus contributing to algorithmic transparency, clarity with respect to news ranking processes and uses of user-generated data' (Exposure Draft Explanatory Material 5).[10]

---

[8] Digital Advertising Services Inquiry (2020), ACCC. Available at: https://www.accc.gov.au/system/files/Digital%20Advertising%20Services%20Inquiry%20-%20Interim%20report.pdf

[9] Treasury Laws Amendment (News Media and Digital Platforms Mandatory Bargaining Code) Bill 2020, Exposure Draft, ACCC. Available at: https://www.accc.gov.au/system/files/Exposure%20Draft%20Bill%20-%20TREASURY%20LAWS%20AMENDENT%20%28NEWS%20MEDIA%20AND%20DIGITAL%20PLATFORMS%20MANDATORY%20BARGAINING%20CODE%29%20BILL%202020.pdf

[10] Treasury Laws Amendment, Exposure Draft Explanatory Material (1–33), ACCC. Available at: https://www.accc.gov.au/system/files/Exposure%20Draft%20EM%20-%20NEWS%20MEDIA%20AND%20DIGITAL%20PLATFORMS%20MANDATORY%20BARGAINING%20CODE%20BILL%202020.pdf

Google's response had initially been to threaten the withdrawal of their search services in Australia, apply pressure from the US government and appeal to Google users in Australia. However the Australian government is aware of Google negotiating a settlement in France in the context of a similar issue. Google had relented in France, where they had to abide by a strict new copyright law and agreed to pay publishers for news content after talks between Google France and the Alliance de la Presse d'Information General (ALAI) although it had removed its Google New Service in Spain after a controversy over payments.[11] Google, has, under its News and Discover service, launched a $1 billion fund (Google News Showcase) to pay for content including the use of snippets to 'Germany's Der Spiegel, Frankfurter Allgemeine Zeitung, Die Zeit, Tagesspiegel and Rheinische Post, Australia's Schwartz Media, the Conversation, Private Media and Solstice Media, and Brazil's Diarios Associados' (Google to start paying for content 2020). On 5 February 2021, Google announced the launching of its News Showcase in Australia, agreements with seven news publishers and 25 mastheads including 25 mastheads, including *The Canberra Times, The Illawarra Mercury, The Saturday Paper,* and *Crikey* (Taylor 2021),[12] thereby signalling its intent to remain in Australia and negotiate with publishers although on its own terms and independently of the Bargaining Code and the government (Packham 2021).[13] It remains to be seen whether this is acceptable to the Australian government and whether Google and Facebook will relent on some of other recommendations including the need for algorithmic transparency and clarity with respect to their uses of user data. It would, in any case, have been politically naïve for Google to pull out of Australia since there has been a change in the mindset of a number of governments who have communicated their intent to regulate the large platforms. On 18 February 2021, Facebook announced its intent to disallow the sharing of 'news' on its

---

[11] Google to start paying some news publishers for content (2020), Associated Press, 25 June. Available at: https://apnews.com/article/633b21eefd458c71fdddbadc9653b58d

[12] Taylor, J. (2021), Google launches News Showcase in Australia, The Guardian, 5 February. Available at: https://www.theguardian.com/technology/2021/feb/05/google-launches-news-showcase-in-australia-in-sign-of-compromise-over-media-code

[13] Packham, C. (2021), Google opens paid-for Australia news platform in drive to undercut government's content payment law, Financial Post, 5 February. Available at: https://financialpost.com/technology/google-opens-paid-for-australia-news-platform-in-drive-to-undercut-governments-content-payment-law

platform—an action that could backfire on it in the light of strong cross-party political support to regulate Big Tech in Australia.

The ACCC Report: Key Recommendations

The ACCC's report is important for a number of reasons. (1) It is among the first, comprehensive attempts by a government to signal its intent to regulate platforms in the interest of competition, transparency, and based on an in-depth understanding of the business that is based on ad revenues. (2) The 23 recommendations, as for instance, 19 on privacy and the need for informed consent, highlight limitations within the Privacy Act and provide the impetus to update this act in the light of new challenges posed by these platforms along with the need to investigate new areas such as data portability, and the reach and the intricate business models and entanglements of tech services and advertising agencies. It is clear that the ACCC's recommendations have, to some extent been shaped by the General Data Protection Regulation (GDPR) and its emphasis on the rights of 'data subjects'. (3) It also makes a case for platform literacies in the community (12) and in schools (13) thus expanding the remit of media education and an early start to new digital natives and their understanding of algorithmic manipulation, fake news, disinformation, and informed consent—thus enabling them to engage with devices, platforms, and content. This recommendation needs to be seen in conjunction with the fact that one of the goals of the Australian school Curriculum is media literacy—'creative and critical thinking, and exploring perspectives in media as producers and consumers' (Australian Curriculum)[14] making it quite unique when compared to school curricula in other parts of the world (Notley & Dezuanni 2019).[15] There have been key reports on media literacy in Australia that point to the fact that teachers surveyed recognize the need for students to have critical digital literacy skills including their ability to engage with and negotiate digitally

---

[14] Australian Curriculum, Aims, Media Arts. Assessment and Reporting Authority (ACCRA). Available at: https://www.australiancurriculum.edu.au/f-10-curriculum/the-arts/media-arts/aims/
[15] Notley, T. & Dezuanni, M. (2019), Advancing children's news media literacy: Learning from the practices and experiences of young Australians (689–701), Media, Culture & Society, 41 (5).

available news and the need for the Australian Curriculum to include a most consistent and updated approach to the teaching of media literacy (see Nettlefold & Williams 2018).[16] A more comprehensive survey to date of teachers in Australia by Dezuanni, Notley, and Corser (2020:24)[17] includes the following observations and rationale for investments in media literacy 'The 2020 Covid-19 pandemic has highlighted the role and importance of news media literacy in Australia. At critical junctures in the development of the outbreak both misinformation (false information not designed to create harm) and disinformation (false information designed to create harm) has been spread throughout Australia, by local, international and sometimes unknown sources. The reasons people fall for false claims are many and complicated. In many cases those who create disinformation are very sophisticated in their ability to manipulate and mislead'. (4) The report makes a clear case that the compensation should also be made available to the two public broadcasters, the ABC and the SBS in contrast to the position of the government that indicated that as public broadcasters, they fall outside of the advertising-based revenue model and thus do not qualify. Nevertheless, the fact that the ACCC had recommended 'stable and adequate funding' (9) for the two public broadcasters remains important in a context in which both the Liberal government and private broadcasters such as News Corp have relentlessly attacked the ideals of public interest journalism espoused by these two public broadcasters. (5) Recommendation 6 encourages a harmonized media regulatory framework with implications for the Australian Communications and Media Authority (ACMA), the peak body involved in media regulation. And (6) the recommendations also include the need for a specialist digital platforms branch (4) within the ACCC to monitor platforms.

[16] Nettlefold, J. & Williams, K. (2018), Insight Five: A Snapshot of Media Literacy in Australian Schools (1–16), Institute for the Study of Social Change, University of Tasmania, Hobart. Available at: https://www.utas.edu.au/__data/assets/pdf_file/0005/1144409/Insight-Five-Media-Literacy.pdf

[17] Dezuanni, M., Notley, T., & Corser, K. (2020), News Literacy and Australian Teachers: How News Media Is Taught in the Classroom (1–32), The Digital Media Research Centre, QUT & The Institute for Culture & Society, Western Sydney University, Sydney. Available at: https://www.westernsydney.edu.au/__data/assets/pdf_file/0012/1689447/Teaching_Media_Literacy_web_version.pdf

The government immediately accepted six of the recommendations and will consider ten more. The key recommendations fully supported by the government include the following:

- 'Investing $26.9 million in a new special unit in the ACCC to monitor and report on the state of competition and consumer protection in digital platform markets, taking enforcement action as necessary, and undertaking inquiries as directed by the Treasurer, starting with the supply of online advertising and ad-tech services.
- Commencing a staged process to reform media regulation towards a platform-neutral regulatory framework covering both online and offline delivery of media content to Australian consumers.
- Addressing bargaining power imbalances between digital platforms and news media businesses by asking the ACCC to work with the relevant parties to develop and implement voluntary codes to address these concerns. The ACCC will provide a progress report to Government on the code negotiations in May 2020, with the code to be finalised no later than November 2020. If an agreement is not forthcoming, the Government will develop alternative options which may include the creation of a mandatory code.
- Conducting a review of the Privacy Act and ensuring privacy settings empower consumers, protect their data and best serve the Australian economy, which builds on our commitment to increase penalties and introduce a binding social media and online platforms privacy code announced in the 2019–20 Budget' (Response to Digital Platforms Enquiry 2019).[18]

To date, there have been few academic writings on the ACCC's Digital report with the only exception being Terry Flew and Derek Wilding's (2020)[19] article in the journal *Media, Culture & Society* 'The turn to

[18] Response to Digital Platform Inquiry, (2019), The Hon. Josh Frydenberg MP. Treasurer of the Commonwealth of Australia, December 12. Available at: https://ministers.treasury.gov.au/ministers/josh-frydenberg-2018/media-releases/response-digital-platforms-inquiry

[19] Flew, T. & Wilding, D. (2020), The turn to regulation in digital communications: The ACCC's digital platforms inquiry and Australian media policy (48–65), Media, Culture & Society, 43 (1).

regulation in digital communication: the ACCC digital platforms inquiry and Australian media policy'. This article provides a history to the
ACCC's report and engages with issues related to the market dominance
of these platforms and its impact on the news business and public interest
journalism in particular. They argue that ACCC's inquiry has displaced
the parallel hearing on media concentration and the need for media diversity, highlight the difficulties faced by small population countries to
subdue global media platforms and issues related to national sovereignty
and deal with the complex nature of regulation of both legacy and digital
media. While this article does raise some significant issues, I will in the
following pages deal with two of the recommendations made by the
ACCC that have been accepted by the Australian government and that is
indicative of a willingness to bring regulation in line with requirements in
the era of platforms and an economy that is fuelled by the monetization of
consumer data sets.

Specialized Monitoring Unit

The Australian government invested A$27 million in the establish-
ment of the Digital Platforms Branch at the ACCC in December 2019.
This unit will contribute to the Digital Services Inquiry that is scheduled to be completed on 25 March 2025 and will have the power to both
monitor and enforce. Two ongoing projects are a five-year monitoring
of anti-competitive behaviour by digital platforms in Australia along
with competition in ad tech and online advertising services. The Digital
Advertising Services Inquiry will investigate markets for the supply of
digital advertising and in particular look into the concentration of power
in these markets, consumer harms, and transparency in the auction and
bidding processes:

- 'competition in the markets for the supply of digital platform services, particularly considering the concentration of market power, the
  behaviours of participants, barriers to entry and an analysis of the
  services offered by suppliers;
- consumer harm that is derived from the practices of suppliers of
  digital platform services; and

- local and overseas digital platform market developments' (Barber & Edwards 2020).[20]

While there has been an attempt to push back on the ACCC's inquiry from industry and scholars in Australia (associated with RMIT) and linked to the Oregon-based International Centre for Law & Economics who have even questioned the ability of ACCC's specialized branch to identify market failure and shape digital markets and contribute to market distortions (Allen et al. 2019),[21] it is clear that such contributions are out of step with the general sentiment in favour of regulating Big Tech platforms. Another example of moves to regulate platforms that is very similar to what the ACCC has tried to do is the UK Government's Competition Markets Authority (CMA) that in 2020 recommended setting up a Digital Markets Unit (DMU) within the CMA mandated to investigate firms that have Strategic Market Status (SMS) (Perica, Sergie, & Waters 2020).[22] The CMA has released an interim report on the digital advertising market in the United Kingdom and recommendations that include the establishment of a code of conduct for platforms funded by digital advertising, establish a Digital Markets Unit within the CMA with enforcement powers and the power to make significant interventions: 'a. Data-related interventions (including consumer control over data, interoperability, data access and data separation powers) b. Consumer choice and default interventions c. Separation interventions' (Online Platforms and Digital Advertising Report, (CMA 2020:34).[23] The CMA's response was shaped by the Furman Report (2019:5),[24] *Unlocking Digital*

---

[20] Barber, S. & Edwards, M. (2020), Digital platforms branch tasked with investigating the digital advertising services industry in Australia, Lexology, Bird & Bird, 19 February. Available at: https://www.lexology.com/library/detail.aspx?g=9d6daf4b-63ce-43e3-95db-6029e0dd89af

[21] Allen, D., Auer, D., Berg, C., Hurwitz, J., lane, A., Manne, G. A., Morris, J., & Potts, J. (2019), Submission on the final report of the Australian Competition and Consumer Commissions Digital Platforms Inquiry (1–24), International Centre for Law and Economics, Portland, Oregon. Available at: https://www.newswire.com/news/scholars-challenge-conclusions-of-accc-digital-platforms-inquiry-20995675

[22] Perica, N., Serie, P., &Waters, P. (2020), So you want a new approach to regulating digital platforms: Too easy (not), Gilbert & Tobin, 12 December. Available at: https://www.gtlaw.com.au/insights/so-you-want-new-approach-regulating-digital-platforms-too-easy-not

[23] Online Platforms and Digital Advertising, Market Study Final Report (2020:1–437), Competitions & Market Authority, 1 July. Available at: https://assets.publishing.service.gov.uk/media/5fa557668fa8f5788db46efc/Final_report_Digital_ALT_TEXT.pdf

[24] Unlocking Digital Competition: Report of the Digital Market Expert Panel (2019), Furman Report. Available at: https://assets.publishing.service.gov.uk/government/uploads/system/uploads/attachment_data/file/785547/unlocking_digital_competition_furman_review_web.pdf

Competition that, among a number of recommendations, established a case for a Digital Markets Unit within the CMA—'the establishment of a digital markets unit, given a remit to use tools and frameworks that will support greater competition and consumer choice in digital markets, and backed by new powers in legislation to ensure they are effective'.

In the EU, regulation is carried out both at a regional level and at a national level in the 28 states that belong to the EU. While the EU's Observatory on the Online Platform Economy monitors core principles related to competition and to the welfare of data subjects as foregrounded in the EU's Digital Services Act (DSA) and Digital Markets Act (DMA) including the role played by core platform providers as gatekeepers role and that use their power over entire platform ecosystems to act as private rule makers and engage in anti-competitive behaviour, there are multiple bodies involved in platform regulation. However, core regulations related to the DSA and DMA will be carried out at a regional level by the European Commission, a prospect that Big Tech might prefer over negotiating with member countries whose data protection laws could be a lot more stringent. The GDPR mandates the position of Data Protection Officers both in public bodies such as the EU and private organizations and there are National Data Protection Authorities in every EU State, who jointly make up the membership of the European Data Protection Board. At the level of EU member states such as in France, there is the Commission Nationale de l'Informatique et des Libertes (CNIL) while in Germany, each of its 16 states have their own data supervision authorities and data protection commissioners.

Privacy

Mann et al. (2018:323–324)[25] offer a rather telling inditement of the state of Privacy in Australia in an article in the *International Communication Gazette*. 'Australia is unique as the only liberal democracy that does not have a comprehensive set of human rights in its Constitution (like the

---

[25] Mann, M., Daly, A., Wilson, M., & Suzor, N. (2018), The limits of (digital) constitutionalism: Exploring the privacy-security (im)balance in Australia (369–384), The International Communication Gazette, 80 (4).

US) or a legislated Bill of Rights (like neighbouring New Zealand) at the federal level. Of the few rights that do receive constitutional protection in Australia, privacy and individual security are not among them, and free expression receives only limited protection via the implied right to political communication ... At the state and territory level in Australia, the Australian Capital Territory (ACT) and Victoria both have human rights legislation which introduces individual rights including privacy and personal security. However, the enforcement mechanisms for these bills of rights are weak: courts cannot invalidate laws for a lack of compliance with the enumerated rights ... The lack of enforceable protections leaves many groups vulnerable to human rights violations and without any means of redress ... There are various areas of concern where the Australian government is violating international human rights standards, particularly in relation to refugees ... and Indigenous/First Nations peoples'.

In Australia, there is a Privacy Act (1988) administered by the Office of the Australian Information Commissioner (OAIC) that offers guidelines to the ways in which federal agencies and private bodies should handle personal information. There are 13 Australian Privacy Principles (APPs) within the Act including privacy regulations for private health care providers along with surveillance legislation at a federal level. However, small business operators with less than $3 million annual turnover are exempt from this Act as are political representatives at federal and State levels, contractors and subcontractors of political parties, and volunteers for political parties (Valles 2018).[26] The major political parties in Australia with the exception of the Greens have continued to support their exemption from privacy laws even when they use the personal information of citizens in their electoral campaigns—a stance that defies reason in the light of the Cambridge Analytica scandal and misuse of the personal data of Australian citizens. (Munro 2018).[27] Some states in Australia such

---

[26] Valle, D. (2018), Australia should strengthen its privacy laws and remove exemptions for politicians, The Conversation, 22 March. Available at: https://theconversation.com/australia-should-strengthen-its-privacy-laws-and-remove-exemptions-for-politicians-93717

[27] Munro, K. (2018), Australia's major parties defend privacy exemption over Cambridge Analytica, The Guardian, 22 March. Available at: https://www.theguardian.com/australia-news/2018/mar/22/australias-political-parties-defend-privacy-exemption-in-wake-of-cambridge-analytica

as Queensland and Tasmania have their own privacy legislations (see Privacy in Your State)[28] and this would suggest the reality of a 'patch-work' of laws related to privacy and of issues arising thereof. Privacy is also covered under other legislations including the Healthcare Identifiers Act (2010), Telecommunication Act (1997), Government Data Matching Act (1990), and Crimes Act (1914), among other Acts. (For a comprehensive listing of privacy and surveillance laws, see the site Privacy and Surveillance: EFF.)[29] While there was an attempt in 2008 under the Australian Law Reform Commission to review privacy law, the Federal Attorney General Christian Porter instituted a review of the Privacy Act in 2020 including the exemptions enjoyed by small business and political parties. A key issue with privacy in Australia is that citizens do not have the right to sue for any breach of personal data. Their complaint has to be channelled through the OAIC, an organization that is under-resourced with a poor record for taking enforcement actions (Goggin et al. 2019).[30] The ACCC's Digital Platform Inquiry has certainly contributed to privacy becoming a public concern. Most notably, it has raised the issue of privacy as a consumer right applicable to all data subjects and highlighted the need for a uniform privacy law that is up-to-date and that is a response to the challenges faced by citizens living data-driven lives over which they have little control. Among the more general concerns are the following:

- 'the scope of application of the Privacy Act, in particular the definition of "personal information" and current private sector exemptions
- whether the Privacy Act provides an effective framework for promoting good privacy practices
- whether individuals should have a direct right to sue for a breach of privacy obligations under the Privacy Act

---

[28] Privacy in Your State, Office of the Australian Information Commissioner. Available at: https://www.oaic.gov.au/privacy/privacy-in-your-state/
[29] Privacy and Surveillance, Electronic Frontiers Australia. Available at: https://www.efa.org.au/privacy/
[30] Goggin, G., Vromen, A., Weatherall, K., Martin, F., & Sunman, L. (2019), Data and digital rights: Recent Australian developments (1–19), Internet Policy Review, 8 (1).

- whether a statutory tort for serious invasions of privacy should be introduced into Australian law, allowing Australians to go to court if their privacy is invaded
- whether the enforcement powers of the Privacy Commissioner should be strengthened' (Witzleb 2020).[31]

A further concern with the Privacy Act is with its epistemic authority and juridical scope in the light of the CDR that gives consumers rights over data portability in relation to personal data collected in the banking, energy, and telecommunications sectors. The CDR is meant to stimulate competitiveness in these sectors by giving consumers the right to own their data and help them to compare and switch services. The ACCC is the chief regulator of the CDR and has been at the forefront of investigating privacy breaches including a case against Google for not disclosing changes to their privacy policy and the way that consumer data is collected and used. The case specifically related to Google's acquisition of DoubleClick, a company that offered ad tech services to publishers and advertisers. Google had initially indicated that it would not combine user data from DoubleClick although it changed its privacy policy in 2016 in the light of combining both data sets. ACCC has argued that Google's attempt to seek consent from its users for this change was flawed and was not based on transparent, consent-seeking processes (Downes 2020).[32] Burdon and Mackie (2020) in an article on the CDR highlight the potential issues with privacy in Australia in the light of two laws, the Privacy Act and the CDR, the former is based on a regulatory mechanism that is principles-based and the latter that is prescriptive (233). While the principles-based Privacy Act offers the leeway to organizations to interpret and implement privacy, the fact that the OAIC has had a chequered record with its 'enforcement' does suggest limits to traditional approaches to privacy in Australia. In contrast, the ACCC's approach to privacy is aggressively on the side of consumers—an approach

---

[31] Witzleb, N. (2020), So, 83% of Australians want tougher privacy laws. Now's your chance to tell the government what you want, Lens, 17 November. Available at: https://lens.monash.edu/@politics-society/2020/11/17/1381694/83-of-australians-want-tougher-privacy-laws

[32] Downes, J. McMahon, F., Griffin, J., Rodrigues, S., Kayis, D., & Tan, E. (2020), The ACCC's new case on Google's collection and use of consumer personal information, Allens, 3 August. Available at: https://www.allens.com.au/insights-news/insights/2020/08/accc-vs-google-on-col lection-and-use-of-consumer-personal-information/

that has been shaped by a recognition of the primacy of user data in the data economy on the one hand and the need to regulate data controllers on the other. However, as Burdon and Mackie (2020:235)[33] have argued, the CDR, despite its forward-looking nature, is not based on privacy as a foundational principle but is linked to the need for data portability. They highlight the fact that without a 'core jurisprudential understanding of what privacy means ... that two tracks of judicial understanding could emerge'. The moves to review the Privacy Act will enable a more comprehensive understanding of privacy principles in the light of the challenges posed by data privacy. Nevertheless, and despite the ACCC's relative lack of experience dealing with privacy, it has demonstrated a keen appetite to regulate Big Tech companies in the context of the compulsions of privacy.

Working in a country that has failed to demonstrate any urgency to reform or regulate the media in the past, the ACCC's Digital Inquiry has been a significant moment. Although its primary aim to make Big Tech compensate the news industry that has suffered from the migration of ad revenues online remains a complex issue, its scope has been sufficiently wide and its recommendations broad enough to deal with issues related to competition in digital markets and the rights of data subjects including most importantly, that of privacy.

---

[33] Burdon. M. & Mackie, T. (2020), Australia's consumer data right and the uncertain role of information privacy law (222–235), International Data Privacy Law, 10 (3).

7

Platform Regulation

Challenges and Opportunities

When both Republicans and Democrats in the United States agree to regulate Big Tech, albeit for different reasons, it points to the fact that there is unease across the major political divides in the United States over Big Tech's power to shape the spread of ideas, set the boundaries for political and cultural discourses and the terms for economic exchange. At the heart of this unease is the potential for Big Tech to manipulate the affective, emotional behaviours of data subjects on a range of issues from voting to dating. In the United States, Big Tech has, in particular, helped untether the First Amendment right of Free Speech from any grounding in rights and responsibilities. On the one hand, platforms such as Facebook have enabled connectivities and voice although on the other hand, and with the benefit of hindsight, we now recognize the pitfalls of unmoderated, unrestrained speech, the sub-cultures that it begets and the consequences of cross-platform viralities that can result in the un-hindered circulations of fake news, misinformation, disinformation, to heightened anxieties and violence off and online. In other words, the 'all speech' is okay dictum has led to what has been a very disruptive attack on facts, truth, and truth telling. We do of course recognize the fact that there are many sides to a story and that certain truths favour the status quo. Be that as it may, in the offline world, as in the world of legacy media, there are codes, legal instruments, and everyday practices that determine the boundaries and limits of media reporting and breaches that invite censure. That frameworks for checks and balances have been disrupted by the social media mantra that all speech is good. The disruption that this has caused is both widespread and deep-rooted and the most pro-found loss is reflected in our inability to shape understandings and create a consensus on the issues that affect us profoundly—the political choices

Platform Regulation. Pradip Ninan Thomas, Oxford University Press. © Pradip Ninan Thomas 2023.
DOI: 10.1093/oso/9780192887962.003.0007

that we make and our vision of a good society. The key issue here is that the meanings of words like rights and freedom are being shaped by Big Tech. Our inherited meanings of rights and freedoms have been displaced by these new meanings that are enticing but that are not based on foundational values.

As the information scholar Prof. Robin Mansell (2019:7)[1] in an unpublished paper describes the background to this regulatory moment: 'the combination of changes in the digital world, and in the everyday offline world, are destabilising what have come to be more or-less taken for granted norms concerning how human beings get on together; how social relations are constituted; and how trust and respect for others are fostered in Western societies … We may look back from the future and see that regulatory inaction, or ineffective regulation, has helped to foster incivility, inequality, and dissension in our societies. This will happen if decisions to encourage participation in a digital ecology are led predominantly by a collective fascination with the potentialities of data and data-processing technology. If, in contrast, such decision are led by the norms and controls consistent with information systems that sustain human beings, and their dignity, there may be a chance to sustain an inclusive and beneficial social order'. There is a sense in which Lawrence Lessig's (1999:89)[2] descriptors of the four regulatory constraints that impinge on behaviour in cyberspace—law, norms, markets, and architecture—are themselves the target for regulators of Big Tech today since the objective is to redesign a regulatory environment fit for the challenges posed by Big Tech across multiple sectors. The regulation of Big Tec via the GDPR, for example, is attempting to redraw the terms of *Code*, to make it more transparent and accountable to principles that are related to the public good. The EU is legislating asymmetric measures and more stringent obligations and sanctions for Big Tech companies, thus obviating the potential for greater harm. It is clear though that despite the protestations from Mark Zuckerberg, the CEO of Facebook, in favour of regulation, Big Tech companies are investing in the stalling, subversion, and disabuse of

[1] Mansell, R. (2019), Digital Platform Regulatory Challenges—Is History Repeating Itself? (1–8), Department of Media and Communications, London School of Economics and Political Science. Available at: http://eprints.lse.ac.uk/102467/1/Mansell_digital_platform_regulatory_challenges_accepted.pdf

[2] Lessig, L. (1999), Code and Other Laws of Cyberspace, Basic Books, New York.

regulation through co-option and direct investments in projects that are meant to highlight their *kosher* motives. An example of 'ethics washing' is Facebook's $7.5 million dollar grant to the Technical University of Munich to set up a new Artificial Intelligence ethics centre (Mathews 2019).[3] Other scholars such as Alpert (2020:1250)[4] are sceptical of the individual-oriented data privacy framework enshrined in the California Consumer Privacy Act (CCPA), which has been hailed as a progressive piece of legislation. He has argued in favour of substantive regulation replacing the 'disaggregated, decontextualized, and atomised data dumps if society is to assert more democratic control over the "common good" of personal data'. Alpert's concern highlights the very general issue of a regulatory free for all, as countries scramble to deal with the challenges posed by Big Tech in the absence of any guiding frameworks—such as that provided by the OECD for the taxation of Big Tech. A multitude of jurisdiction-specific regulations is not the answer since perceived regulatory failure can be used as an excuse to make a case for minimum or no regulation.

While the chapters in this book have dealt with regulation from the perspectives of competition, taxation, and privacy, it is important to highlight the fact that the scope of regulation is vast precisely because Big Tech disruptions have affected the hearts and minds of consumers, productive processes, the terms of economic and cultural exchange, and our valuation of goods and services. Big Tech is forcing a re-regulation. The mainstreaming of 'surveillance' by the State and Big Tech adds yet another level of complexity. While regulatory instruments such as the GDPR have framed rules for Big Tech's consent-seeking from data subjects, the State has escaped the same levels of scrutiny regarding their uses and abuses of surveillance technologies and the vast powers that they exercise in using citizen surveillance as a way to monitor, deter, discipline, and punish. So while we rightly focus on Big Tech, how privacy is affected by the State is a key issue that has to be addressed in the context of the real and imagined threats to national security. Furthermore, there is also the need to regulate the conflicts of interest that arise when the interests of the State

---

[3] Mathews, D. (2019), Big tech funding AT ethics research to 'delay and avoid' regulation, Times Higher Education, 25 April 2019, Issue 2405. Available at: file:///Users/uqpthom4/Downloads/ProQuestDocuments-2022-08-29.pdf

[4] Alpert, D. (2020), Beyond request and respond (1215–1254), Columbia Law Review, 10 (5).

and Big Tech become so enmeshed that it is difficult to see the difference between e-government and e-government by Big Tech. A key question for us to answer is—how can we, in a world that is increasingly run by artificial intelligence, machine learning, and algorithmic oversight, ensure that norms, rules, and values take precedence over, subordinate, and frame the workings of Big Tech operations and cultures? At the same time, there is an absolute need for independent bodies to oversee transparency and accountability in State surveillance activities that have expanded across the world and that are not answerable to anyone.

The recent tilt towards the regulation of Big Tech is a direct consequence of a number of scandals involving Big Tech and evidence of their omissions and commissions. These include the Snowdon revelations, the Cambridge Analytica scandal, Big Tech collusions connected to Brexit, the 2016 elections in the United States, multiple data breaches across a number of public and private sector organizations in both the developed and developing world and evidence of their anti-competitive behaviour in jurisdictions around the world. Regulation, in other words, has largely been a reactive response to events and exposes that have revealed the extent of data misuse and the abuse and complicity of Big Tech in dubious forms of transnational data management. While there is no denying the fact that recent attempts to regulate Big Tech contributes to the public good, one can argue that there is a need for regulation to begin from a larger reckoning with data as a primary basis for lives lived in multiple data worlds.

In a volume on the relationship between technology and law—Law 3.0, Roger Broadsword (2020:60)[5] alerts us to the shaping of legal doctrine and regulatory thinking by technology. He describes three registers in which regulators communicate their regulatory intent to regulatees.

1 in the moral register, the coding signals that some act, x, categorically ought or ought not to be done relative to standards of right action— regulators thus signal to regulatees that x is, or is not, the right thing to do;
2 in the prudential register, the coding signals that some act, x, ought or ought not to be done relative to the prudential interests of

---

[5] Broadsword, R. (2020), Law 3.0: Rules, Regulation & Technology, Routledge, London.

regulatees—regulators thus signal to regulatees that x is, or is not, in their (regulatees') self-interest;

and

3 in the register of practicability or possibility, the environment is designed in such a way that it is not reasonably practicable (or even possible) to do some act, x—in which case, regulatees reason, not that x ought not to be done but that x cannot be done.

Applying this reasoning to the regulation of Big Tech, it is clear that both the moral (public good) and prudential (self-interest) registers were ignored in an environment that was favourable to self-regulation and to the regulatory enablement of the scale, operations, expansion, and capillary embrace of Big Tech. In fact, the basic strategies often used by the State to regulate—to command (law used to pursue objectives), deploy wealth (contracts, subsidies to affect behaviour), harness markets, inform, act directly (to contain an issue), to confer protected rights (via the construction of rights and liabilities) (see Baldwin, Cave, & Lodge 2011:2)[6] were, for some inexplicable reasons, kept in abeyance. Arguably, Big Tech did take advantage of the tax cuts, subsidies, and other types of preferential treatment that came their way and that was brokered by the State. A string of scandals including privacy breaches along with global evidence of their anti-competitive behaviours and disruptive influence on democratic norms, standards, and rules resulted in regulators taking the decision to draw boundaries and set the terms for the operations of Big Tech. While there is the ever-present danger that the regulation of Big Tech is offset by the rapid obsolescence of regulation in the context of technological innovation, disruption, and dissonance, that prospect has not deterred the State and regulatory bodies from investing in the regulation of Big Tech.

The State, Civil Society, and Regulation

Colin Scott (2004:161–166)[7] had, in a very thoughtful and critical chapter on regulation and the post-regulatory state, argued that in the

---

[6] Baldwin, R., Cave, M., & Lodge, M. (2011), Understanding Regulation: Theory, Strategy & Practice, Oxford University Press, Oxford, New York.

[7] Scott, C. (2004), Regulation in the age of governance: The rise of the post-regulatory state (145–174) in Jordana, J. & Levi-Faur, D. (Eds.), The Politics of Regulation: Institutions and

post-regulatory state, the power to regulate is dispersed among numerous bodies from supra-national to national. He highlights the variety of regulatory norms and rules including norms and rules for self-regulating regimes (autopoiesis), the variety of control mechanisms, variety in controllers, and variety in controlees in a post-regulatory state. It would seem that in the environment for contemporary platform regulation—the big issues such as competition and privacy have become core aspects of the regulatory state while platforms are simultaneously being urged to design systems that are secure and in the interests of both business and the public. Self-regulation, in other words, is no longer the only possibility but is part of a larger package of regulation involving a variety of stakeholders including the State and civil society. While there are significant civil society actors in the platform regulation space including networks such as the JustNetCoalition,[8] which is a formidable network consisting of organizations and individuals in civil society involved in advocating for larger justice issues related to the Internet, the extent to which their contributions are making a considered difference in the regulation and policy arenas remains unclear. There is, however, good reason to believe that civil society coalitions in Europe have a better prospect to contribute to the discussion in the EU than in other regions around the world. A report on a civil society platform regulation meeting—Civil Society Recommendations on how the EU should regulate online platforms, Democracy Reporting International (2020) that was held in Germany and funded by the German government, included a number of civil society organizations including 'Stiftung Neue Verantwortung (SNV), AlgorithmWatch, The Portuguese University ISCTE-IUL MediaLab, Democracy Reporting International (DRI) Access Now, Avaaz, Civil Liberties Union for Europe, Civitates, Election Observation and Democracy Support, EU DisinfoLab, European Digital Rights, European Partnership for Democracy, German Marshall Fund, Institute for Strategic Dialogue, Open Society European Policy Institute, Panoptykon, and Tactical Tech' (Democracy Reporting International

---

Regulatory Reforms for the Age of Governance, Edward Elgar Publishing Ltd., Cheltenham, UK and Northampton, Massachusetts.

[8] JustNetCoalition. Available at: https://justnetcoalition.org/

2020).[9] Another meeting also in the EU, organized by the NGO AlgorithmWatch: Governing Platforms Project recommended that the EU establish a new institution explicitly linked to investigating transparency issues linked to the Digital Services Act 'The institution might also explore the possibility of engaging the broader European public in the development of research agendas (see e.g. lessons from the Dutch National Research Agenda 16) or by incubating pilot projects that explore the possibility of connecting users and researchers through fiduciary models. Independent centers of expertise on AI/ADM at national level, as proposed by AlgorithmWatch and Access Now17, could play a key role in this regard and support building the capacity of existing regulators, government and industry bodies' (Governing Platforms, Final Recommendations 2020:4).[10] AlgorithmWatch's mandate related to Algorithmic Decision Making (ADM) includes an explanation of why transparency is required:

1. 'ADM is never neutral.
2. The creator of ADM is responsible for its results. ADM is created not only by its designer.
3. ADM has to be intelligible in order to be held accountable to democratic control.
4. Democratic societies have the duty to achieve intelligibility of ADM with a mix of technologies, regulation, and suitable oversight institutions.
5. We have to decide how much of our freedom we allow ADM to preempt' (AlgorithmWatch).[11]

The regulation of Big Tech is a chapter that still is in its infancy, although as the case studies in this book reveal, there is an appetite and mood in favour of regulation across the world. In the United Kingdom,

[9] Democracy International Reporting (1–15), (2020). Available at: https://democracyreport ing.s3.eu-central-1.amazonaws.com/images/21972020-07_CS_Tech_Regulation_Recommen dations.pdf
[10] Governing Platforms Project (2020), Putting Meaningful Transparency at the Heart of the Digital Services Act Why Data Access for Research Matters & How We Can Make It Happen (1–6). Available at: https://algorithmwatch.org/wp-content/uploads/2020/10/Governing-Platfo rms_DSA-Recommendations.pdf
[11] What we do, AlgorithmWatch. Available at: https://algorithmwatch.org/en/what-we-do/

Germany, France, China, South Korea, and Japan, among other countries, there have been attempts to file anti-trust suits against Big Tech and/or to pass legislations that are explicitly directed towards creating a level playing field and strengthening competition and privacy standards. The National Diet of Japan enacted the Improvement of Transparency and Fairness in Trading on Specified Digital Platforms Act on 27 May 2020 and that is aimed at strengthening disclosure obligations and effective redress (Amato & Maezawa 2020).[12] In South Korea, the competition chief Joh Sing-wook who is in charge of the FairTrade Commission has vigorously followed up on data monopolies and privacy protection (CPI 2019).[13] And in China, Jack Ma's Alibaba and financial services Ant group came under government scrutiny just as a planned $30 billion IPO was in process. The Cybersecurity Law (2016) along with the Personal Data Protection Law (2020) and the Civil Code of the People's Republic of China (2020) offer privacy protections (Greeven 2021).[14]

What to Regulate in the Context of Data Ubiquity

However, the issue of who to regulate and what, and when and how to regulate remains complex not least because 'data' is not just the fuel for Big Tech but is common currency in the transactional, global economy that we are all a part of. Financial data in the banking sector, insurance data in the insurance sector and health data in the health sector, data in the biotech sector, the use of personal data in the political sector and in electoral politics—all these have implications for privacy and point to the overwhelming reality of data as common currency across multiple sectors. That presents us with a quandary—can one regulate Big Tech without at first getting to grips with the primacy of data as common currency today

---

[12] Amato, J. J. & Maezawa, T. (2021), Japan: Japanese legislature passes act to regulate Big Tech platforms, Mondaq, 21 January. Available at: https://www.mondaq.com/antitrust-eu-competit ion-/1024456/japanese-legislature-passes-act-to-regulate-big-tech-platforms

[13] South Korea: Tech giants worry over new antitrust enforcer, Competition Policy International, 5 September. Available at: https://www.competitionpolicyinternational.com/ south-korea-tech-giants-worry-over-new-antitrust-enforcer/

[14] Greeven, M. (2021), China is pushing to control big tech, with e-commerce giants coming under the microscope, The Fashion Law, 11 January. Available at: https://www.thefashionlaw. com/china-is-pushing-to-control-big-tech-with-e-commerce-giants-coming-under-the-mic roscope/

and as an increasingly significant basis as input and output for all manner of economic productivity? In other words, where does one begin from? I am not suggesting that the only way is to begin from a comprehensive understanding of data since there are, as the maxim goes, different ways to skin a cat. The right to be forgotten and data portability highlights the complexities of personal data and attempts to provide some certainty to the dignity of personal data in a world where the mining of affect has become the basis for a mode of production that is essentially unregulated. These are ideas and concepts in their own right that have the potential to result in actions that protect one's right to be visible or for that matter invisible, untethered from information grids and from 'presence' in online socialities. However, there needs to be attempts to understand data as a building block, as common currency in national and global contexts, so that in the least, there are multiple conversations on what can and cannot become datafied. In other words, is it inevitable that all information is destined to become data or is it the case that a case can and should be made for some areas to be cordoned off from the predations of data colonization? In other words, some types of information that cannot and should not be datafied and commodified? The questions raised highlight a pressing issue—that governments and regulatory bodies are only very gradually building up knowledge of the complexities and intricacies of datafication and of the ways in which Big Tech fall short in their professed commitment to the public good. As van Dijk, Poell, and de Waal (2018:21)[15] in a chapter on governance in their book *The Platform Society*, observe:

most national legislatures simply *do not have a fitting vocabulary* to capture the socio-technical finesses of an evolving ecosystem of platforms that threatens to undermine many established societal arrangements. They also *lack a refined taxonomy* of techno-commercial mechanisms that can adequately delineate power relationships between various actors. Data flows, path dependency, sector-agnostic algorithms, vertical and horizontal lock-ins, active users as currency for value

---

[15] Van Dijk. J., Poell, T., & de Waal, M. (2018), The Platform Society, Oxford University Press, New York.

accretion, and search engine degradation ... for instance, are not part of a common legal discourse (emp. authors).

In the absence of such a vocabulary, it would seem that regulation reflects the most salient, arguably predictive trajectories such as the exploration of a Google Tax and data privacy. These issues are by no means easily resolved since establishing an action plan to do with the taxation of a transactional economy will be difficult to implement as is the case of data privacy in the context of the multijurisdictional nature of data flows. It remains to be seen as to whether the OECD/G20's agreement to levy a minimum 15% tax rate on global MNCs earning more than US$950 million in any calendar year that is scheduled to begin in 2023, will be universally enforced. Another issue that has come to the fore in the context of protectionism and data nationalism is data localization—an issue that the EU has explored in the context of its plan to establish a Digital Single Market and in India, where localization is tied to contesting the perceived data imperialism practiced by Big Tech. This stance was reflected in India not signing on to the G20's Data Free Flow with Trust charter linked to the Osaka Track agreement—free flow being a traditional moniker invoked by those who subscribe to the thesis of media imperialism and that describes the reality of constrained, over-determined, one-way flows from the West to the Rest (Chaudhuri 2020).[16]

Data Regulation and Politics

In India, data localization has been mandated by the Reserve Bank of India and the SriKrishna Commission; the latter has recommended a raft of measures including that 'critical personal data' can only be processed in data centres located in India (Punj 2020).[17] Basu, Hickok, and Chawla

---

[16] Chaudhuri, R. (2020), India and the geopolitics of technology, The Hindustan Times, 4 December. Available at: https://www.hindustantimes.com/analysis/india-and-the-geopolitics-of-technology/story-dd2DGyVvuhH7XmOfVSpLwL.html

[17] Punj, S. (2020), India's data tussle with big tech, India Today, 3 December. Available at: https://www.indiatoday.in/india-today-insight/story/india-s-data-tussle-with-big-tech-1624622-2019-12-03

(2019:28)[18] from the Bengaluru-based Centre for Internet & Society have written a comprehensive report on data localization in India. They allude to the contested nature of data localization in India and highlight the reasons for local conglomerates such as Reliance Jio supporting data localization. Reliance does have the economic resources and market power to absorb the costs of data localization and invest in data centres although the authors also point out that mere localization is no guarantee for innovation. The rhetorical significance of Reliance's commitment to data localization is highlighted in this paragraph taken from their Integrated Annual Report (2018–2019:92)[19] 'Data Localisation—Jio has been a strong supporter of local storage of data, which is critical for national interest and security given the increasing sophistication of cyber-attacks. Data localisation will also spur investment in creating server and cloud capacity in India, incentivising research and development and creating employment in line with the Government of India's "Make in India" initiative. Jio believes that Indians are the true owners of their data and the ownership should not be transferred to any corporate entity. Without the consent of the user, no data should be collected, processed or used by any corporate. This would require a regulatory framework to ensure that corporates are taking adequate measures to ensure data protection. Reliance is not averse to partnerships with Big Tech who are often against data localization. In November 2020, the Competition Commission of India (CCI) approved Google investing $4.5 billion and a 7.7% stake in India's largest telecommunications platform—the Reliance-owned Jio Platform (Singh 2020).[20] The CCI had also approved, in June 2020, Facebook's $5.7 billion investment in a 9.9% stake in Jio Platforms (Singh 2020).[21]

---

[18] Basu, A., Hickok, E., & Chawla, A. S. (2019), The Localisation Gambit: Unpacking Policy Measures for Sovereign Control of Data in India (1–90), Centre for Internet & Society, Bengaluru. Available at: https://cis-india.org/internet-governance/resources/the-localisation-gambit.pdf

[19] Integrated Annual report, 2018–2019 (1–460). Reliance Industries Limited. Available at: https://www.ril.com/getattachment/0461b91d-61ce-44d3-8a8c-3918a9b32ff7/AnnualReport_2018-19.aspx

[20] Singh, M. (2020), India approves Google's $4.5 billion deal with Reliance' Jio Platforms, TechCrunch, 11 November. Available at: https://techcrunch.com/2020/11/11/india-approves-googles-4-5-billion-deal-with-reliances-jio-platforms/

[21] Singh, M. (2020), India approves Facebook's $5.7 billion deal with Reliance Jio, TechCrunch, 24 June. Available at: https://techcrunch.com/2020/06/24/india-approves-facebooks-5-7-billion-deal-with-reliance-jio-platforms/

The story of Reliance does suggest that the regulation of Big Tech in India is bound to remain a difficult task. Data localization that favours local corporates over 'foreign' players is not a solution precisely because of a lack of clarity regarding data privacy.

There is also the critical issue to do with the beneficiaries of regulation. In whose interests is regulation advocated and justified? In the case of the Australian Competition and Consumer Commission's (ACCC) Digital Inquiry, the main beneficiary is legacy media, in particular the duopoly of News Corp and Nine Entertainment who together dominate the media landscape in Australia. As I have previously noted, New Corp owns 70% of news output in Australia and its outlets including the Australian and Sky News are highly partisan and support the Liberal Party and conservative politics. In fact like its counterpart in the USA Fox News, Sky News in Australia, long after the Biden administration had been sworn in the United States, continued to peddle claims that the election had been stolen. In India, it is clear that the Equalisation Levy has helped the government earn income from taxing the platforms involved in an advertising-based business model. However, the brand of data nationalism adopted in India clearly favours the leading corporate house in India, Reliance, which is close to the ruling government and has clearly been involved in anti-competitive behaviour in the mobile phone and e-commerce markets. In other words, the professed commitments to a 'level playing field', although rhetorically progressive is not an easily actionable item given the reality of network externalities and the reach enjoyed by incumbent players in this space. While Microsoft's Bing, which has a little more than 3% of the search market in Australia, has expressed its desire to increase its market share if Google decides to withdraw, that, in itself, is not a solution given that this might only result in another monopoly provider in the search space.

Platforms as Public Utilities

The solutions resulting from regulation can be opaque, are often politically motivated and may result in indirect support for competitors jockeying for position in an already narrow field. So the question that remains and needs to be grappled with is how can regulation contribute to the public good and support public interests? The critical theorist of the digital Mark Andrejevich (2013:130)[22] has argued for 'real' alternatives and public investments in platforms that are freed from the commercial imperatives that constrain Big Tech. 'Why not consider the possibility of a public service social networking platform—one freed from the commercial imperatives that require Facebook to engage in detailed tracking practices that greatly expand its infrastructure needs, which in turn require it to more aggressively 'monetise' its user base? Why not imagine the possibility of a well-crafted public service search engine whose algorithms are driven by the goal of creating a more informed citizenry, rather than one more likely to click on advertising links or visit commercial sites?'. Andrejevich's questions highlight issues with the limits of platform moderation that is based on an algorithm-based selection dynamic that favours those who pay. There are other scholars such as the lawyer Lina Khan (2017:713)[23] presently the Chair of the FTC in the United States who, in an article on Amazon, explored the possibility for its regulation as a public utility. She describes the critical role that it plays as an intermediary and the power it exerts, as a 'retailer and as a marketing platform, a delivery and logistics network, a payment service, a credit lender, an auction house, a major book publisher, a producer of television and films, a fashion designer, a hardware manufacturer, and a leading provider of cloud server space and computing power'. It is their extensive business portfolios based on horizontal and vertical integration and ability to use predatory pricing and anti-competitive practices to control the terms of exchange that has led to numerous anti-trust suits in the United States. Khan, while taking into account Amazon's extraordinary power in e-commerce to control the terms of sale by sellers and producers, suggests that 'non-discrimination', a principle that sits at the core of the theory of public utilities, could be the basis for the reinvention of Amazon as a public utility. 'One approach would apply public utility regulations to all of Amazon's business that serve other businesses. Another would require breaking up of Amazon and applying non-discrimination principles separately; so, for example, to Amazon Marketplace and Amazon Web Services as distinct entities' (800). Khan has continued her argument in

---

[22] Andrejevich, M. (2013), Public service media utilities: Rethinking search engines and social networking as public goods (123–132), Media International Australia, 146 (1).

[23] Khan, L. M. (2017), Amazon's antitrust paradox (710–805), The Yale Law Review 126 (3).

favour of regulating Big Tech and in a 2019 article has made a case for recovering the need for structural separateness when confronted with dominant intermediaries such as Google, Amazon, and other Big Tech companies (1091).[24] The overwhelming power of Big Tech to control critical 'chokepoints'—be it in the areas of online payments, online search, or the nature of freedom of expression—highlights the unprecedented might of a handful of global MNCs.

There are, however, those who contest the public utilities model and prefer that the market is left to deal with and sort out platform concentrations and their anti-competitive behaviour. Adam Thierer (2013:297),[25] for example, reflects this school of thought when he suggests that 'It is essential to have a little faith in the entrepreneurial spirit and the dynamic nature of markets built upon code, which have the uncanny ability to evolve and upend incumbent "tech titans" seemingly every few years'. Given contrary evidence on the ability of the market to correct itself, Thierer's argument can best be applied to a 'perfect' market, an ideal that is reflected in books although not in the real world.

While the recent flurry of legislations across the world aimed at curbing the power of Big Tech is welcome and there are signs that some ideas such as the right to be forgotten and data portability may have universally cache, an issue of concern is whether what is being generated today in different jurisdictions amounts to a hotch-potch of ideas on regulation and whether, in this context, there is a need for the equivalent of a MacBride Report for the 21st century that investigates the world's problems with data and data controllers? The MacBride Report 'Many Voices, One World' (1980)[26] was a report on the world's communication deficits and was published by UNESCO as a response to the call for a New World Information and Communication Order (NWICO). In other words, the need for principles that have universal validity generated through multistakeholder processes and that accounts for the fact that there are different data ecologies, some that require regulation in the

---

[24] Khan, L. M. (2019), The separation of platforms and commerce (973–1098), Columbia Law Review 119 (4).

[25] Thierer, A. (2013), The perils of classifying social media platforms as public utilities (249–297), CommLaw Conspectus, 21.

[26] MacBride, S. (1980), Many Voices, One World: Towards a New More Just and More Efficient World Information and Communication Order, International Commission for the Study of Communication Problems, UNESCO, New York.

interests of the public good while others—such as in the context of in-formal economies that need to be recognized for the role that they play in democratizing access to data and contributing to data education. Is there a need for some standardization and harmonization of data regulation and who should be involved in such processes? UN bodies, governments, civil society, the private sector? Are there lessons to be learned from multistakeholder processes such as at the Internet Consortium for the Assigned Names and Numbers (ICANN) and the Internet Governance Forum linked Dynamic Coalition on Platform Responsibility (DCPR)[27] or do these experiences provide a good enough reason not to embark on multistakeholder-based models? What are the data ecologies in local contexts and what can be done to map key actors who can be involved in establishing understandings, frameworks, and projections for data jour-neys in national contexts?

The real consequence of data colonization certainly is a key reason for the regulation of Big Tech. The fact that Twitter and Facebook exer-cised their power to both stifle and silence the President of the United States (with good reason, I may add) in the aftermath of the storming of Capitol Hill, Washington in January 2021 after enjoying years of profiting from his online activities, and Google and Facebook threats to curb, even withdraw from Australia if the 'Bargaining Code' became law, reflects the power that they can and do exercise. The fact that Facebook had tem-porarily banned Australian users from sharing and viewing all Australian and international news content, even government-run health and emer-gency websites, points to their unwillingness to 'pay' for news and to make an example of Australia to other countries such as Canada that is also considering the regulation of Big Tech (Facebook blocks Australian users 2021).[28] In some ways, Big Tech has no choice but to function in uncharted waters in which rules and principles such as in areas such as hate speech and censorship are not a given but are responses that are rela-tive to the countervailing power of authorities such as the EU and the US government. Google can threaten the Australian government although

---

[27] Dynamic coalition on platform responsibility, Internet Governance Forum. Available at: https://www.intgovforum.org/multilingual/content/dynamic-coalition-on-platform-res ponsibility-dcpr
[28] Facebook blocks Australian users from viewing or sharing news (2021), BBC, 18 February. Available at: https://www.bbc.com/news/world-australia-56099523

it simply has to negotiate with the EU and hope that in the United States, their lobbyists, anti-antitrust economists, and pro-Silicon Valley media will enable them to keep functioning as monopolies. In countries such as India, where data nationalism is one aspect of a complex relationship with Big Tech characterized by cartographic anxieties with Google Maps, close affinities between Google India and the ruling Bharatiya Janata Party's IT wing, periodic shutdowns of social media, and censorship in 'problem' areas such as Kashmir and the farmer's agitation in Delhi, Big Tech simply has to toe the government line even if they do occasionally protest. The complicity between Big Tech and governments is an issue that simply has to be investigated.

I think that one of the key issues in the context of ensuring competition and privacy is the absolute need for data literacy. The fact that most people online are relatively unconcerned with their 'data journeys' has provided the perfect opportunities for Big Tech to 'manage' and monetize data generated by data subjects. The fact that the ACCC's report has recommended media education for the digital era in schools is critical to the education and shaping of digital literacies that enable the data-evaluation and data use skills of data subjects for whom data generation is an essential practice of connected, daily life. It is important that students learn to distinguish fake news from real news and truth from misinformation and engage with news sources that can be trusted. Admittedly, this is a difficult task in a context in which legacy too media are either in retreat or are controlled by the government or by private corporations and fact checking contested. I would argue that media education should not focus exclusively on the critical reading of texts but should also engage with the political economy of Big Tech, with data ecologies, with ideas such as data as a public good, Big Tech as public utilities, and the right to be forgotten. In other words, data education should be an opportunity for students to engage with critical ideas and aspirational goals for both Big Tech and data subjects.

Let me conclude with how I began. Platform regulation is of recent origin. It is unfinished business. And there are myriad complexities. For example, apart from Weibo and WeChat, that are based in China, the world's major digital platforms are of US origin. This remains an issue given that successive US governments have, despite bipartisan attempts to rein in Big Tech, protected the global interests of their MNCs. These

MNCs have a significant economic presence in jurisdictions around the world and their profits are generated on a global basis although many of these BigTech firms have not decentralized their operations to any significant extent. The parallels with the governance of the Internet are just too obvious. Just as the International Consortium for Assigned Names and Numbers (ICANN), the peak body involved in the governance of the Internet is instituted in California and until recently functioned on the basis of a Memorandum of Understanding with the US Department of Commerce, the world's major platforms are based in the United States, and are indeed protected by US law. While it is the case that there is continuing bipartisan support for the revocation of Article 230, Communications Decency Act (1996) that gave platforms immunity from lawsuits emerging from third-party content on their sites that violated copyright, material on the sexual exploitation of minors, etc., the US government typically supports and defends the global economic interests of their platforms. One major issue arising from global platform breaches and evidence of platform collisions with foreign interests in the gaming of US elections is whether the US government will resort to regulation in the interests of multilateral solutions rather than a strictly national one. There are numerous attempts to self-regulate although irrespective of US-government solutions, governments around the world now recognize the need for hybrid approaches to regulation—self-regulation along with legislated regulation. To make matters more complex, traditional rule makers such as the International Telecommunications Union and the World Bank are investing in/making claims for their involvement in the regulation of platforms. While it certainly makes sense to think through domain-specific platform regulation, as for example, ITU's involvement in spectrum allocation, civil society really must be involved in access and justice issues thereby contributing to a radical pluralism in platform regulation (Cammaerts & Mansell 2020).[29] As I was in the process of writing the final words in this volume, the Australian mediasphere was buzzing with the news of Facebook's decision to block the sharing of local and international content by users and media companies in light of the new Media Bargaining law that mandates Google and Facebook

[29] Cammaerts, B. & Mansell, R. (2020), Digital platform policy and regulation: Towards a radical democratic turn (135–154), International Journal of Communications, 14.

to pay for news on their sites. Google has decided to make its own deals with media players in Australia while Facebook has decided to block any content that is deemed to be news. Is this a case of Facebook flexing its muscle and making an example of a recalcitrant Australia? Or is it a case of a global corporation's overreach and hubris? It is clear that the stakes are high. In all probability, Facebook will go back to the negotiating table since the political fallout from unilateral, blocking actions may not result in a desirable outcome. Given all that has happened in this space, the scandals involving Big Data, the anti-competitive behaviour of Platforms, algorithmic political polarization, and privacy breaches, I do not seegovernment's relenting although they too are heavily invested in platforms and the platform economy. The key issue though is whether all these varied regulatory actions will benefit the common good. That process of affirming data as a common good has already begun. On 25 November 2020, the European Commission published a draft of a legal framework—Data Governance Regulation that is meant to contribute to a functioning Digital Single Market. It has three objectives—(1) The *sharing of public sector data*, (2) the establishment of a new, exclusive, neutral, bodies—*data sharing service providers* who will be tasked with the data exchange of personal and non-personal data, and (3) encouraging *data altruism* and the voluntary consent for the use of privately held data for the common good (Evans, Modrall, & Sinclair 2020).[30] The OECD/G20's ratification of a globally applicable minimum taxation regime (15%) for MNCs in October/November 2021 can also be seen as a step in the right direction. The report of the Payments System's Review submitted by Scott Farrell to the Treasury, Australia in 2021[31] explicitly deals with the need to regulate FinTech in the interests of Australian citizens and end claims for 'platform exceptionalism'. Haggart and Tusikov (2021)[32] in an article on the need to regulate social media by the Canadian

---

[30] Evans, M., Modrall, J., & Sinclair, M. (2020), EU data governance regulation—a wave of digital, regulatory and antitrust reform begins—Part 1. Norton Rose Fulbright, Blog Network, 17 December. Available at: https://www.dataprotectionreport.com/2020/12/eu-data-governance-regulation-a-wave-of-digital-regulatory-and-antitrust-reform-begins-part-1/

[31] Payments System Review: From System to Ecosystem (2021), Submitted by Scott Farrell, Commonwealth of Australia, June (1–115). Available at: https://treasury.gov.au/sites/default/files/2021-08/p2021-198587.pdf

[32] Haggart, B. & Tusikov, N. (2021), Regulating the digital economy in three parts: The conclusion, Centre for International Governance Innovations, June. Available at: https://www.cigionline.org/articles/beyond-speech-regulating-the-digital-economy-part-three/

government, highlight a key recommendation made in the report by Scott Farrell. 'Platform exceptionalism contends that services delivered online or via an app should be treated differently than the same services delivered offline (see Uber and taxis, or Airbnb and hotels). The Farrell Report calls on Australian regulators to set rules based on the nature of the service, not on the entity providing the service. Under this rule, platforms providing payment services would not be treated differently than traditional financial institutions offering the same services. Simply put, the claim of "platform" would no longer be a perceived or actual regulatory advantage'.

Admittedly, these are small steps. And perhaps there are those who will argue that the EU's invocation of the public good and the right to be forgotten is limited to EU citizens. However, and despite such misgivings, these are important principles that have the potential to become universally applicable and the basis for norms and rules necessary for the regulation of Big Tech.

Index

For the benefit of digital users, indexed terms that span two pages (e.g., 52–53) may, on occasion, appear on only one of those pages.